Praise for *Invisible Learning:*

"For over fifteen years, my former co-a⸱

has been teaching statistics successfully at the Harvard Kennedy School. In these challenging times, it is more important than ever that change-makers understand both the opportunities, and the limitations, of the data they rely on. Now, thanks to David Franklin's remarkable book, anyone can be a fly on the wall in the class that persuaded generations of future leaders to love statistics."

- Michael Kremer, Winner of the 2019 Nobel Prize for Economics and Professor at the University of Chicago

"In my 11-year tenure as the Dean of the Harvard Kennedy School, I heard so frequently from faculty and students alike about the extraordinary teaching that Dan Levy consistently did that I asked him to lead a major new teaching initiative for the entire school. This wonderful new book by David Franklin reveals the magic behind his class.

I was transfixed. I knew Dan and his collaborators were good, I just did not know they were THIS good. If you want to understand what propels spectacular teaching, read this book. If you want to understand how and why statistics matters to us all, read this book. And good gracious, if you are ever expected to teach or learn statistics at any level, devour this book."

- David Ellwood, Dean of the Harvard Kennedy School (2004-15) and advisor to President Bill Clinton

"In nearly 20 years working with Harvard faculty to advance effective teaching, I have never had the pleasure to work with a more dedicated and effective educator than Dan Levy. While it may be tempting to assume that Dan was simply born a great teacher, the truth is that teaching is a skill that must be learned. Dan's open-mindedness and dedication to continuous improvement keep him fresh and engaged, and judging by the student evaluations of his teaching, he's a stand-out success year after year. The lessons in this book from Dan's thoughtful, creative, and collaborative approach to teaching should be required reading for educators everywhere.

- Carolyn Wood, Director of Educational Quality Improvement at Harvard Medical School

"In my 37 years of teaching leadership at Harvard, I have used student cases and the dynamics inside the classroom as a living set of cases from which to examine properties of leadership, authority, and collective problem-solving.

In this book, David Franklin deftly uses leadership frameworks to analyze what happens in a very different kind of course: API-209, an extraordinary statistics course taught for many years by my colleague Dan Levy at the Harvard Kennedy School. If you are interested in discovering how to the design and deliver courses that aim to generate collective responsibility and collaborative learning, I highly recommend you read this book."

- Ron Heifetz, Senior Lecturer in Public Leadership at the Harvard Kennedy School and author of Leadership Without Easy Answers

"Being a student in the class that inspired this book and an enthusiast of teaching and learning, I was delighted to learn that David wrote this book. David was a very dedicated course assistant of mine and we had several conversations about pedagogical strategies. I was always impressed how carefully he perceived each movement in the class and tried to understand how it would affect the learning process. He was also a great teacher. In my preparation for an exam, I clearly remember going to him for help since I know he would not be satisfied until I had a very clear understanding of the concepts. I hope that this book will be as insightful and magical for you as it has been to me whenever I talk about education with David."

- Bia Vasconcellos, student at the Harvard Kennedy School

"As part of Dan's teaching team, I was lucky enough to work closely with him on his innovative approach to teaching and student inclusion. Having spent hours talking with David how to make Dan's class even better, I know he is in a perfect position to write about it. This book shares all the secrets of our teaching team and our class. It's wonderful to know that people outside Harvard will now have access to Dan's remarkable pedagogy."

- Sophie Gardiner, former course assistant for Dan Levy

INVISIBLE LEARNING

The magic behind Dan Levy's legendary

Harvard statistics course

David Franklin

First published in 2020 via Kindle Direct Publishing

Copyright © 2020 by David Franklin

Foreword copyright © 2020 by Dan Levy

10 9 8 7 6 5 4 3 2 1

The right of David Franklin to be identified as the author of this work has been asserted by him in accordance with the Copyright, Designs and Patents Act 1988.

ISBN 979-8-5708322-6-8

Cover design by Ana Laura Urquiza Gomes

www.invisible-learning.com

For Gan and Boompa, who taught me how to learn,

and for Catri, from whom I learn so much

CONTENTS

Foreword

David Franklin is a student you never forget as a teacher. In the Fall of 2018, he was my student in API-209, a course I teach at the Harvard Kennedy School. He should not have been my student. As a brilliant mathematician from the University of Cambridge, he knew all the statistics concepts that my course was designed to teach. But because my course was required for the Masters in Public Administration program he was enrolled in, he took the class just like everyone else.

As expected, he finished at the top of the class. But contrary to what I expected, he engaged in the course energetically and enthusiastically, with the same curiosity as someone who was learning the material for the first time. I did not realize at the time that part of what had him so engaged is a passion for learning and a knack for helping others learn.

The following year, I was fortunate to have David serve with distinction as a teaching assistant for the course. At the same time, he took leadership courses at the Kennedy School and learned important concepts and ideas developed by my legendary faculty colleague Ron Heifetz. I have huge respect for Ronnie both as a teacher and as a leadership scholar and practitioner, but I had not seen much of a connection between his ideas of leadership and my teaching practices – until David came along.

As the semester progressed, David interpreted what he was observing in my classroom through his perspective as a student interested in how people learn, as well as some of the frameworks he was learning in his leadership courses. He would share his observations with me through written notes after every class, and I would be fascinated by the insights he was giving me on a course I had taught for over 15 years. One day in the middle of the semester, I was telling Catri Greppi, another teaching assistant for the course and David's partner, how useful I was finding David's notes. She said, in her usual direct matter-of-fact way, "David should write a book about your course", to which I quickly replied: "And who would be interested in that?!" The next day, David was in my office and the book you are now holding in your hands was born.

Given that the book is centered on a course I teach and the way I teach it, you might be wondering if I am in an objective position to convince you about its value. I am certainly in no position to write about whether my teaching and that of many colleagues who directly contributed to the teaching approaches highlighted in this book (including but not limited to Jonathan Borck, John Friedman, and Teddy Svoronos) is worth highlighting in the way David has done.

But here is what I can say: During my over 20 years of teaching, I have devoted countless hours to reflecting on my teaching. One of my sacred practices is to write reflections after every class session in which I meticulously document what worked and what didn't. I base these reflections on my own observations, feedback from teaching assistants, data collected from students, time spent on each part of the class, and other sources.

I thought I had a good sense of what happened in my classroom and why I did the things I did. David's book has made me see what I do in the classroom and the craft of teaching more generally in a completely new light. He has given me insights and language to understand the teaching practices I employ, and in the process, he has helped me become a better teacher. For this alone, I am eternally grateful to him.

David also brings another valuable perspective to this book. That of a student. Most teaching and learning books are written by faculty members, teachers, teaching coaches or teaching professionals. David is none of these. Although he read some seminal work in teaching and learning to ground his writing, he does not claim to be a teaching expert. But his perspective is valuable nonetheless because it is grounded on the experience of the learner, and because he put many of his ideas to the test when he was trying to help his fellow students learn. I understand that his sister is an exceptional teacher in England, and I suspect her ideas are also coming through in David's writing.

Who is this book for, you might ask? After all, it is a book that brings together the craft of teaching, the field of statistics, and the practice of leadership. If you are interested in any of these areas, I think the book can provide you with some valuable ideas. If you are a professor or teacher, the book will help you see your craft in a new light regardless of whether you have the time or interest to use in your own teaching the approaches that David highlights in this book. If you are interested in statistics, the book will provide you with ways of applying the discipline to the world at large. If you are a leader or interested in leadership, you might see applications of the ideas of adaptive leadership beyond your world.

Finally, if you are interested in two or more of these areas and how they can connect with each other, I believe you will find this book profoundly stimulating and a true delight to read.

Dan Levy
Cambridge, MA
November 2020

Introduction

In Cambridge, Massachusetts, the Harvard Kennedy School sits a few yards to the north of a bend in the Charles River. The school is a hub for idealism and liberal thought: a thousand promising students, from a hundred different countries, find their way there every year with the hope of changing the world for the better. All of them, in different ways, have been identified as 'future leaders' in their countries. The school aims to teach them the skills they will need to go into government and make change in a sea of politics, or to work with non-governmental organizations and charities to maximise their influence on the world. Named after John F. Kennedy, its motto is his most famous line: "ask not what your country can do for you, but what you can do for your country."

The students who come in are passionate and motivated. They are doers, not thinkers. On their first day, they are given the class list. They can learn about leadership, geopolitics, climate change, negotiation skills, lobbying, energy, trade, race relations, intersectional feminism. There is a class called 'Getting Things Done'. World-famous figures like Samantha Power, Larry Summers, and Dani Rodrik teach there to starry-eyed students. Presidents and Prime Ministers make regular visits to the beautiful, cylindrical forum at its entrance on John F. Kennedy Street. The school is a place not for theorists, but for practitioners.

As a result, it fills many with dread and disappointment to learn that one of their core courses in the first year is named 'API-209: Advanced Quantitative Methods'. This is exactly the sort of class they came here to *avoid*. They are not here to do a math degree: they want to learn about the world. Nervously, they shuffle in on Day One anxious that their lack of expertise will be exposed. Many start to wonder whether coming to the Harvard Kennedy School was really the right idea in the first place. At the front of the class, Professor Dan Levy, who has been awake in nervous anticipation to prepare since five o'clock in the morning, ensures all his tech is working and prepares to introduce the course.

This is a book about Dan's class. Three months and twenty-seven lectures later, many of these students have tears in their eyes as it ends. The class, more commonly known as just 'Statistics', has been a transformative experience. They have learned new ways of looking at the world, while being part of a community in which they feel safe. They feel *seen* as themselves. Old fears of mathematics, so deeply entrenched through years of difficult experiences, have softened. The tears come from the loss of that environment.

The group may be full of idealists, but they are a harsh lot when it comes to rating their professors. Many are paying a lot of money to come to Harvard and expect the quality of their teaching to reflect the prestige of the institution. One professor, who arrived with a wealth of elite-level experience in creating change around the world, made a great effort to get to know his students and was well-liked by everyone in the class. But his voice was a little soft, and the course – a new addition to the syllabus – was a little unstructured. He was rated a disappointing 3.2 out of 5 and left Harvard after just a year. The students didn't appreciate being the 'guinea pigs' for a flawed course.

Another came in with a top reputation as an economist in his home country, and was visibly excited to be teaching at Harvard. But he kept running over time, and he did not do enough to translate equations into words. Students did not like having mathematics thrown at them without real world applications: that is not what they came to the Kennedy School for. He scored just 2.1 out of 5, and once again, did not return to Harvard the following year. These are not 'Uber' ratings in which you get five stars unless something goes wrong. Anything above 4.2 is a good rating, and above 4.5 is exceptional.

Dan's instructor rating has been above 4.8 out of 5 for five years in a row. For a class which students are forced to take, in a subject usually thought of as dry and difficult, this is unheard of. Many students say that Dan is 'different' to other professors. Faculty at the Kennedy School have said he is 'special', and joked that 'we can't all be Dan'. But no-one can articulate exactly what he does to stand out. To many, there appears to be something magic about it.

There is no such thing as magic, and this kind of talk makes me suspicious. The myth of the 'brilliant man' is commonly used to justify an excess of trust in male competence. Humans have an innate desire to anoint messiahs to give them the answers to complex questions. Dan is *not* special, but he is devoted. When others put his success down to just 'being Dan', they are avoiding the work of understanding and explaining that success. They are also underestimating the amount of hard work that goes into the learning environment he creates.

This is a personal story in which I try to answer the question of why Dan's teaching is successful. I spent two years in his class – the first as a student, and the second as part of the teaching team. During that second year, I noted down everything he did in the classroom to inform this book. I wanted to know what, *specifically*, he did to generate such a bond with his students – and how he managed to teach them something about statistics that stuck with them.

It is important to emphasise that Dan is not unique. There are thousands of teachers around the world who are equally successful with similar – or better – methods. The ideas he uses are commonplace across many elementary and high school classrooms, even if at a university they appear innovative. The 'magic' idea of giving responsibility back to the students for their own learning is met with a shrug by schoolteachers: "of course we do that". Many of the concepts in this book have been known in the education literature for some time, and in common use in schools across the world. That said, I hope the framework developed in this book will be of value, or at least of interest, to anyone with a passion for pedagogy.

Central to this framework will be the concept of 'invisible learning': the harsh reality that true understanding of an idea is hard to see and hard to measure. Learning depends to a great extent on the strength of the bonds between the members of the classroom, but these bonds, too, are invisible. In Dan's class, his goal of maximising learning is shared by the students: in a sense, *everyone* is on the teaching team. Students take responsibility for their own learning, and for their own teaching. The effects are invisible, but essential for the work that gets done in the classroom. The idea of student responsibility creates some fascinating parallels with the study of leadership, and we will go into this in depth in Part II.

As a personal story, it is biased. I am too generous to Dan. The premise of the book is that there is excellence in his teaching that should be examined and explained. Someone who thought the opposite would be unlikely to write this book. I have tried to write with a critical eye, and in interviews with Dan and others, I have questioned his methods. It is certain that there are things that Dan could do differently that would further improve the learning of the class, and we will explore what these might be. As a mathematician, I may also be too generous to Dan in failing to challenge explanations that are difficult for those without numerical backgrounds. The reader, free from the personal biases of the author, may be in a better position to be more critical than I have been competent enough to be.

As a tool to help examine Dan's teaching, I will take the reader inside the classroom. Only a lucky few get to experience the joy of API-209: this book gives away the 'secret sauce' behind one of Harvard's most successful courses. There are great teachers all over the world who entrust students with the responsibility for their own learning. As case studies go, though, Dan's classroom is a good place to start.

The book is divided into five parts. Part I takes the reader into Dan's classroom for the first four classes of term: the goal is to make you familiar with the way he teaches. The style is mostly 'fly on the wall', but we will also step back and think about what he is doing, with help from Dan himself, who kindly agreed to numerous hours of interviews for this book. For those I am thankful to him: the book praises his teaching, but it was not his idea. He would probably have preferred to stay out of the limelight.

Once you have got a taste for the class, we then step outside in Part II to examine how the learning happens. The focus is on the invisibility of learning: we cannot see it happen, and we struggle to measure it. The book considers the numerous invisible interactions happening around the classroom. Our goal is to decipher the 'magic' to which people attribute Dan's success. For anyone who has had a great teacher in the past, I hope it helps you to explain *their* success too. For those frustrated by university professors who were well-qualified but never seemed to 'connect' with their classes, I hope we can also explain their failure.

Armed with our new framework, we then go back to the classroom in Part III to listen into the main part of the course. You will learn, as the students do, about how to answer questions with statistics, and how to avoid pitfalls like cherry-picking, which can lead us to be more certain of our conclusions than the data allow. We will also look at the way Dan adapts to feedback from the students during the course, and resists the temptation to prioritise student satisfaction over maximising learning.

In Part IV, we step back outside to consider the invisibility of *teaching*. Having looked at the invisible building blocks of the wider learning environment in Part II, we now zoom in on the invisible dynamics around the teacher. All those who contribute to the class by speaking, including Dan, are authorised to do so by the rest of the group to varying extents. We will go further into this idea of 'authority', and we will examine ways in which the teacher can raise or lower the tension in the classroom to maximise learning. In this section we also look at the data that Dan gathers, and the constant process of self-evaluation and iteration, to make the class better every year.

Finally, we re-enter the classroom in Part V to put everything together. The pace will be quicker as we close out the book, as it would be foolish to try to cover the whole course. We will only have time to jump between a few of the classes whose teaching illustrates the framework well. This book is for people who would, for any reason at all, be excited to sit in on Dan's class: that might be because you are interested in pedagogy, or statistics, or both. It is targeted at a lay audience, but just as those who have taken statistics courses before find great value in Dan's class, I hope this book may be of value too.

PART I: UNCOVERING OUR BIASES

1

CERTAINTY IS AN ILLUSION

The first class is a lightning bolt to students expecting a dry course of mathematical proofs. Plenty assume that statistics is a branch of mathematics. But mathematics deals with certainty: thanks to Pythagoras, we *know* that the square of the longest side of a right-angled triangle is exactly equal to the sum of the squares of the other two sides. Statistics deals primarily with uncertainty. You might have data from a small sample of the population, such as a poll in which a thousand Americans are asked who they will vote for in the next election. The goal of Dan's class is to teach students what they can infer, and what they should be *wary* of inferring, from the data they have.

Many in the class will go out to become policymakers. In the world of government – and in the private sector too – there is a huge market for certainty. We all want someone to tell us which program to fund, or which stock to buy, or whether the economy will rebound quickly in the years after a global pandemic. Often, no-one has these answers in any form more certain than an educated guess. But so long as we are willing to pay for certainty, there will remain those who claim to provide it.

For these 'future leaders' in the classroom, it is more important to understand the *psychology* of certainty. Often, we believe what we want to believe. Dan starts with an example of the 2016 Presidential Election in the United States. He asks the question:

"Relative to what you expected to happen, how surprised were you when you learned that Trump had won the election?"

This is the first use of 'Poll Everywhere', an online tool to allow instructors to quickly poll students and gauge the results. The students are asked to rate their level of surprise on a five-point scale from 'not surprised at all' to 'beyond shock'. Many describe feelings of total shock and numbness: a large majority indicate some level of surprise. He asks them:

"Why were you surprised?"

An American student, well-versed in U.S. politics, puts his hand up. He says he had been following Nate Silver's 'Five Thirty-Eight' website, which had said that Trump only had about a 30% chance of winning the election.

"In a world of uncertainty," the student ventures, "Five Thirty-Eight is a shining light."

This is a revealing comment: websites like Silver's help to release the tension caused by lack of information. Five Thirty-Eight has everything an election enthusiast could dream up to estimate probabilities of outcomes, as well as attractive tables and graphics. We have an innate anxiety at being unable to predict the future, especially when a worrying outcome is possible. Psychologically, we are drawn to a website that feeds our hunger for information to reduce that anxiety. But even if we were to assume their analysis is right, that same hunger causes us to misinterpret it. The brain has conflicting goals: to provide you with information, and to reduce your anxiety about the worrying outcome. If you wanted Clinton to win in 2016, your brain achieved both goals by accepting the 30% figure but telling you it would not happen.

Dan knows that Five Thirty-Eight is a popular website among those interested in U.S. elections, and he has pre-empted the student's response. He shows a screenshot of the website on the day before the election, confirming the student's earlier assertion on the probabilities: Clinton was 71% likely to win, Trump 29%.

"So Trump was 29% likely to win. What does 29% mean?" he asks.

This is an odd question for many: they're doing a Master's degree at Harvard, and they're being asked what a percentage means. The question is aiming at the gut reaction of the brain to the number. Inevitably, students think it might be a trick question. Dan waits a while, and when no-one raises their hands, he breaks away from the example to tell a story:

"You know, when I first started teaching, I was terrified of silence. I thought, 'oh my god, I've got to do something, they're not saying anything'."

The class laughs: he has eased the tension created by the silence.

"The more I taught, the more I realised that silences are important in a class – they give time for people to think. These days, I'm not afraid of silence at all."

After a few seconds, a woman puts up her hand.

"Well, obviously, I knew that it meant there was some chance that he would win. But it was still a shock that the 29% happened."

Dan asks the class whether any of them has ever checked the weather on their phone to see if it's likely to rain while they're outside. Everyone has.

"What do you do if you see there's a 30% chance of rain?"

"Take an umbrella", several students all say at once.

"It sounds like a lot of you are not too surprised if this 30% chance materialises," Dan replies. "You take precautions against it. The rain, if it comes, doesn't shock you, and you have your umbrella. So why was everyone so shocked about Trump?"

There is a short silence. Another student responds:

"I guess... it was just a shocking event. We're not used to someone like him being President."

Here, the student reveals an *availability* bias in which we expect future outcomes to look like what has gone before. As the psychologists Daniel Kahneman and Amos Tversky first outlined in the 1970s, our minds assess probability by the extent to which instances of an event are available in our memory. Since he could not remember an event 'like' Trump's victory, he underestimated the probability.

Availability bias works in the opposite direction too. If you are a soccer fan whose team is 1-0 up in the last minute and concedes a corner, there is likely to be a frantic voice in your head crying "We're doomed: we *always* concede from corners". Only 3% of corners result in goals, but you remember them more when they happen. The fact that goals from corners are abundantly available in our memories leads us to overestimate the probability of conceding.

After pausing, the student continues: "There's also an element of believing something because you want it to happen."

This betrays a second bias in which we believe what we *want* to be true. This is a form of *confirmation bias*, which exacerbates the distorting effect of availability bias. We seek out information that confirms our existing beliefs or desires, and ignore information that refutes them. Those who did not want Trump to win in 2016, and who had never seen a political outcome like his victory, had two reasons to underestimate the probability in their minds, even if in principle they accepted Nate Silver's figure of 29%. It can go the other way when availability of an outcome is high: if you are convinced your team always concedes from corners, then a goal scored will confirm your belief, but a clearance will be quickly forgotten. Your level of conviction, thanks to confirmation bias, will harden over time.

In both cases, availability and confirmation bias work in tandem to increase the size of our processing error. Dan is trying to make students aware of the things that could prevent them from making accurate judgments in the future. In doing so, he wants them to start thinking probabilistically about the world, and never to be taken in by the illusion of certainty.

ELEVATING STUDENTS

He moves on by briefly telling the class that they will learn about three main uses of statistics during the semester. Statistics can be used to *describe* the world, to assess *causal effects* between one thing and another, and to *predict* outcomes. The first might describe the characteristics of poor households in India. The second might help to investigate whether improved access to finance would help lift them out of poverty. The third might estimate how many households are likely to be poor next year, independent of the reasons why.

In advance of the class, he asked students to fill in a quick survey in which he asked them to describe their own experiences with one of these three uses of statistics. He tells the students that he has read their answers, as well as their student profiles, and asks for volunteers to talk about what they wrote. Several hands go up, and he picks Juliana, a student from Brazil, interested in education. She starts to talk about a program she helped to run in Brazil which assessed whether improvements in teacher training had a causal effect on test scores. As she is talking, her words appear on the big screen, in big quotation marks, along with her photo. The class laughs, and as she looks up she realises she has become a celebrity.

Speaking a year later, Juliana still remembers this moment. "I was very surprised!" she says. "I remember thinking, 'This professor read all the student profiles. He's different.' I felt like he cared a lot." One focus of this book will be to explore why it matters that Dan cares about his students: though it might make them feel happy, the link between caring and learning outcomes is not obvious. When we talk about the invisibility of learning in Chapter 5, we will consider how the invisible bonds that are formed in class contribute to the learning that is done.

This is the first of many instances throughout the semester in which Dan will put student quotes and photos on the screen. Implicitly, the message to the students is "your words are worthy of being quoted". It creates amusement every time, but it also strengthens the bonds between the students and the teacher. It is clear to the students that he respects them, and their experience. By elevating their words, he elevates *them*.

THE BIRTHDAY GAME

In the earlier example, students were 'given' the probability of 29% for a Trump victory and invited to reflect on it. The second example, which he will spend the remainder of the class playing with, teaches a slightly different lesson: *humans are bad at estimating probabilities*. He asks the class to answer the following question on Poll Everywhere:

"What is the probability that at least two people in a group of 20 people will have the same birthday?"

He asks the 80 students to respond based on their 'gut feeling'. Again, students are given five options, ranging from 'less than 1%' to 'above 40%'. About half of them believe the true answer is less than 5%, of which plenty go for the 'less than 1%' option. Only 1 in 6 get it right, picking the highest option: it turns out that the true figure is 41%. He invites those people – 13 in total – to stand up.

"I'd like everyone to applaud those who got the answer right. Well done!"

Figure 1.1: perceptions of the probability that at least two people in a group of 20 will share a birthday

Answer	< 1%	1-5%	6-20%	21-40%	> 40%
Students	15	24	20	8	13

Everyone claps, despite the mild theatrics. The students who got it right are still standing, and looking a little sheepish to have been applauded.

"Now – those who got it right. On the count of three, I'd like you to sit down if you'd seen the question before. 1, 2, 3!"

Everyone who was standing sits down. This gets a huge laugh. The students who got the answer wrong suddenly feel much better about their mistake: in a Harvard class, not a single person got it right unless they had seen it before. One student suggests that he thought it was probably going to be a trick question in which the answer was much lower *or* much higher than they expected, which is why he went so low. But *everyone* who guessed in this way went low, rather than high. It suggests that there is some form of systemic bias in play.

Dan shows a table with the probabilities that no-one will share a birthday in groups of different sizes. It turns out that you only need 23 people for it to be more likely to find a shared birthday than not. With 80 people, the size of this classroom, the probability that there will be no shared birthdays is so tiny as to be virtually impossible. All of this is counter-intuitive, and it's a result of how the brain works. For most people, it is not obvious how to solve this question. Availability bias kicks in, and they seek a similar problem they *do* know how to solve.

This will probably be the following easier problem: "in a group of 20 people, what is the probability that at least one of them shares *your* birthday?" You have a small number of people (20) and a much bigger number of possible birthdays (365). Without doing any calculations, the answer is probably low.

But this is not the question: we wanted to know the probability that *any* pair will share a birthday. There are 20 people who could have shared *your* birthday, but in a class of 20, there are 190 different *pairs* of students, any of whom could share a birthday. This is a much bigger number relative to the 365 days in the year, which is why the true answer is above 40%.

Dan is not done with the fun and games.

"This table says that there will almost certainly be at least one pair of you who share a birthday. Let's test that."

There is an excitement in the class: people want to know who shares a birthday with whom, and whether the table is right or not.

"I'd like anyone who was born on March 22nd to stand up."

Two students stand up. They link eyes from across the room, and smile. The room laughs.

"Now, anyone born on June 14th."

Another two stand up. The room laughs again.

"Now, anyone born on August 2nd."

Two more: the class is laughing at each pair, and re-examining their original takes on how low the probability was likely to be. They are learning through experience about human fallibility with estimating probabilities, and at the same time learning something about their classmates.

"How about December 21st?"

This time, *three* students stand up. This gets a big laugh for two reasons: firstly, it is unexpected, and secondly, it helps to ram home the idea that in a class that size (there are about 3,000 pairs in a class of 80), there are likely to be "coincidences" everywhere.

As the laughter subsides, there is just enough time left in class for Dan to remark that his research led him to realise that for Lina, a student from Germany, it is her birthday *today*. He suggests that on the count of three, the class sing her 'Happy Birthday', and the student looks happy and surprised. She gets a roaring, if discordant, rendition of the famous song from the rest of her cohort. The first seventy-five minute class is over, and everyone leaves with a smile on their face.

STRENGTHENING BONDS

A lot is happening below the surface in the first class of the year. Dan spends several hours in advance of the class getting to know his students. He has access to the student profiles which he has asked them to fill out before the class, so he knows about their backgrounds, ambitions, and interesting facts. He also has their answers to the pre-class questionnaire talking about their experiences, which he was able to draw upon with Juliana.

Most professors do not spend this amount of time understanding the backgrounds of their students, and they do not significantly impact the way in which they teach. Dan believes this is the most important part of his preparation:

"Teaching is a human process at its most fundamental level, and you have to make it personal", he says. "At the end of the day, it is simply about helping another human being to learn."

There is an important sense of humility here. Teaching is about *helping* students to learn. The work in learning is with the students themselves. Too many professors, he believes, see teaching and learning as the same thing. For them, the better they teach, the more the students will learn. All their planning therefore relates to their own actions: how they will explain the next tricky concept, and what they will say at what time. A class goes well if they remember to do what they planned, and badly if they fluffed their lines or went off-track.

Lesson planning is important: at times, Dan obsesses over it. But, in his words, "it's not how the PowerPoint looks that matters. The more you can pay attention to *them*, the more successful you'll be, no matter how great your explanations are." He believes that the interaction and engagement with students is what helps them to learn the most. When the lesson is tailored to the experiences of the students, they will learn more. It gives them more responsibility for their own learning, and takes advantages of related connections they have already made. Importantly, it also strengthens the bonds between Dan and the students: they know he is committed to their learning, and they see that he knows who they are. When students see that a teacher is putting the effort in to support their learning on an individual basis, it makes them more likely to reciprocate.

This idea of 'reciprocation' is an important feature of the first class. Dan describes the syllabus as a *contract*. In return for their attention, engagement, and adherence to class norms, teachers will help students to learn. The first class sets the tone for the strength of this contract: it is not hard for anyone to recall examples from their childhood in which this contract between students and teacher was weak. By getting to know the students and making them feel seen and understood, a teacher can improve their odds of student reciprocation.

The opening example of Trump's election win is designed to set the tone for the rest of the class. This is as much a class about the students' *relationship* with statistics as it is about statistics itself. There is no mathematical, or even statistical, work to be done anywhere in this first lecture, but plenty of challenging students on the way they think. Dan's goal, taken from his mentor Richard Zeckhauser, is to get students to *think probabilistically about the world*. This means engaging with the challenge of understanding and accepting what 29% means in the context of a Trump victory, and fighting the brain's natural inclination to see things in binary terms.

It also serves as an entertaining start to the course. When he first started teaching, he used to open with a big-picture summary of the whole semester. But he realised that it was, in his own words, "kind of boring". He needs to hook people's attention early to motivate and inspire them to learn. This is a core course which students are forced to take. The first lecture, and all that follow it, are designed to avoid any student wondering, 'why do I need to learn this stuff?'. The Trump example, given the strong opinions on his presidency among many of the students, tends to elicit emotional reactions and avoid any early boredom.

The birthday example in the latter half of the class is designed not only to get students to question their own intuitions, but also to build a sense of community in the classroom. Something powerful happens when the students themselves are the examples they are working with. If the exercise had been instead to investigate a set of birthdays randomly generated by a computer to find pairs and triples, the level of student engagement would be much less. By making it about the students, Dan can magnify the interest in the topic.

When the students stand up, either because they got the initial answer right or because their birthday has been called out, the other students all peer round to look at them. There is a fascination with who has been selected and why, just as there was when Juliana found herself quoted on the screen. The bonds between the students strengthen as they interact with each other, and learn things about each other. The same exercise done without the physical act of standing up, and the interactions that it causes, would not be as effective.

ESTABLISHING NORMS

Before the Trump example, Dan takes some time at the start to establish norms and expectations, as is common for many professors. This formalises the idea of the contract. Students are expected to arrive in time to be ready to start at 1.15, and to complete problem sets every week. Some of these problem sets will be long, and difficult. The course will be challenging, even if they have taken statistics classes before. Phones, laptops and other electronics are not permitted, except if required for special assistance or when using Poll Everywhere as part of the class.

The tone in this early part is something that Dan struggles with. His personality, and natural style as a lecturer, is to be friendly to his students. He does not want to be feared: as we will discuss later, to be feared as a professor could weaken the learning environment. At the same time, though, it is important that students realise that they are there to do some serious learning. A big part of the teacher's role is to ensure that the students have a sense of responsibility for their work.

Dan admits: "I don't feel good about how I strike the balance." Sometimes, after this opening class, he comes away feeling like he has been too dictatorial in setting expectations. The reality may be the opposite. Dan's inclination to be friendly with his students probably leads him to be too nice in these early norm-setting minutes. The stern voice required to set norms that students will adhere to is one that he is uncomfortable using.

This is also the case later in the term. Dan concedes that he is "not good at enforcing the punctuality contract". The norms that are set by the teacher are, in some sense, less important than how they are enforced. What the teacher does when a norm is violated for the first time is important in setting the tone for the rest of the class.

Dan's problem is that he struggles to find the stern voice required to enforce that norm. He will jokingly ask latecomers: "what's going on, guys?", while shaking his head in mock exasperation. But his tone is one of humour, and it invites laughs from the class. Using humour at this point might be a mistake: as we will discuss in Chapter 13, humour tends to lower the tension in the room. Enforcing norms effectively may require the tension to be raised instead. Dan's use of humour, and creation of strong bonds with his students, pays great dividends elsewhere in the class. Late students are one area where he probably gets the balance wrong.

A different norm, which Dan establishes to great effect, is that *silences are okay*. We heard his short story to the class about how he feared silence when he first started teaching. His approach to silence may surprise some students: more commonly, silence is taken by professors as an indication of weak participation in class. But the impact of normalising silences is profound: when they are 'permitted', they create space for learning to take place. By refusing to step in and provide them with answers, he gives the responsibility of learning back to the students. The additional time students can use to think then has a dramatic role in increasing participation across the class. We will discuss silences at length later in the book.

AIRPORT IDEAS

The phrases 'certainty is an illusion' and 'think probabilistically about the world', both introduced in the first class, are the first two examples of high-level concepts which Dan calls *airport ideas*. He tells the class:

"These are ideas that I want to be so memorable, so deeply entrenched in your brain and how you view the world, that in five years' time – when I see you at an airport – you will be able to explain those ideas to me even if you've forgotten everything else."

The idea of having 'big picture takeaways' from a course is not new: most professors will structure their classes to include them, and if they do not, the students will let them know about it in their evaluations.

Dan takes it a step further by hammering home the 'airport' message at every available opportunity. This is not just a bullet-point list of 'key takeaways' at the end of a lecture. In the classroom, Dan has three big screens at the top. When he reaches an airport idea, he will have the idea in large font on the centre screen, with nothing else to distract from it. The screens either side will have a memorable image. We will see later on in the class that when he reaches the airport idea of 'cherry picking' – using data selectively to manipulate your audience and support your case – the left and right screens have a small girl picking cherries. When people are asked what they remember from the class, everyone remembers her: the brain latches onto images and stories but forgets bullet points quickly.

There is also something powerful in the concept of an 'airport idea'. On the surface, it links with Dan's plea for what students should do if they meet him in an airport in five years' time. But there are hidden suggestions too: the airport carries an association of being high-level, and fits in with the course's aim of empowering the students with the tools they need to effect change. Once the course is over, they will disperse to airports across the world and put these ideas into practice in their home countries.

2

UPDATING YOUR PRIORS

Having got the students' juices flowing in the first class, which is deliberately light on theory, Dan knows he faces a challenge in keeping their attention when they move into the first stages of probability theory in Class 2. His goal is to teach them about Bayes' Theorem, which is at the foundation of conditional probability. It helps you to update your estimate for the probability of something when you get new information.

For example, you might be confident that your team has a 70% chance of winning a soccer match, but then they concede a goal in the first minute. Bayes' Theorem says that if you want to estimate the probability of winning given the new information, you have look at the data in which teams similar to yours conceded a first-minute goal, and see what proportion of those still won. Perhaps your win probability is now only 40%. In more general terms, if you want to know the probability of an event A given new information B, you look at all the times that B happens and find the proportion of them in which A happens as well. Statisticians call this "updating your priors": you start with a prior belief about something (A), you get new information (B), and you update your belief accordingly (A given B).

Dan wants to get across to the class that the ability to update your priors is more than a statistical concept: it is a way of life. The great economist John Maynard Keynes is supposed to have said "when the facts change, I change my mind. What do you do, sir?" But few of us really stop to ask ourselves what new information might change our minds, especially on questions of politics and identity. We fall into the trap of confirmation bias, in which rather than updating our prior beliefs, we mould new information into a form that will confirm them. This psychological flaw can be catastrophic for our ability to infer meaning from statistics.

The key airport idea is the concept of 'being a Bayesian': someone who updates their beliefs when new information comes to light. He puts up a picture on the screen of Thomas Bayes, the eighteenth-century statistician who formalised the idea mathematically. It's not an inspiring portrait: it's a little smudged, and Bayes looks a bit stern. The class laughs. Dan says,

"So… this is Bayes. Maybe you don't aspire to be Bayes, he's a minister from centuries ago. The model I want you to have, though, is this one."

Next to Bayes on the screen appears a motto in white text on a red background. It reads 'Keep Calm and be a Bayesian'. Dan is getting across to the students that being a Bayesian, which they will learn about in this class, is not just something they will *do*, but a whole new way of *being*. If Dan is successful, it will change the way they look at the world.

WHAT DOES MARTIN DO?

Dan then poses a question on Poll Everywhere. He tells the class about Martin, who is analytic, regularly reads the Wall Street Journal, and invests his personal funds in a well-diversified portfolio. He then asks:

"What are the odds that Martin is a finance professor teaching MBAs, as opposed to being a physician?" *

The three possible answers are 'more likely to be a physician', 'roughly the same likelihood', and 'more likely to be a finance professor'. The students answer using their phones, and only Dan can see the results. He raises an eyebrow, and tells the students that it looks like the class will be helpful to them. The students laugh nervously. He promises to return to Martin at the end of the class, keeping them in suspense for now.

MAMMOGRAMS

Before they hear about Martin and his career, the students are introduced to some of the key ideas that arise from Bayes' Theorem using an example from public health. Of the students in the class, who number between 70-80 each year, there are usually one or two medical doctors planning to become leaders in public health, either in the United States or in their home countries. The question of medical diagnosis is one of the best ways to illustrate Bayes' Theorem

---

* The inspiration for this question comes from the psychologist Daniel Kahneman, who asks a similar question in his book *Thinking Fast and Slow.*

intuitively. Tests for diseases are not perfect, so it is helpful to know, if we take a test and it comes back positive, what the likelihood is that we actually have the disease. It is possible that we registered a 'false positive' instead.

Dan starts with a deliberately provocative question: "should women in their forties get regular mammograms?"

For those who are aware of the positive benefits of early detection of breast cancer, and especially for those with experience of breast cancer in their families, this seems like a no-brainer. But this is a course designed to test the students' assumptions and force them to wrestle with the sorts of difficult policy questions that they will face in their future careers. If the mammogram test is not precise enough and generates a lot of false positives (or false negatives, in which patients with cancer are *not* identified), then the money spent on them may be able to save more lives elsewhere.

Dan is aware of the possible emotional triggers around this issue, but also chose this example because it brings home the life-and-death importance of understanding Bayes' Theorem well. He tells the class: "This is a delicate issue, and many of us have personal experiences with this topic. The goal is not to trivialise or dehumanise, but rather to understand how statistics can inform the debate."

Before doing any of the math, Dan asks the students to spend some time jotting down the advantages and disadvantages they can see to the policy proposal. Pedagogically, the key is that the motivation for the statistical exercise is clear before any numbers are introduced. There are always a few mathematicians in the audience, but for most students, unmotivated math during a lecture will cause them to withdraw – as the unfortunate economics professor from our introduction found in his evaluations.

After a discussion about the sorts of things they might expect to trade off against each other when evaluating such a policy, Dan moves to a question he has asked the class to answer online, separately, in advance of the class. The question is based on data available for women aged between 40 and 50.

"Eight of every 1,000 women have breast cancer. Of these 8 women with breast cancer, 7 will have a positive mammogram. Of the remaining 992 women who don't have breast cancer, about 70 will still have a positive mammogram. Imagine a sample of women who have positive mammograms in screening. What percent of these women actually have breast cancer?"

Almost all of those who received this question answered it correctly: these students are smart, and many have seen this kind of thing before. Doing it online before class, they had time to think about their answer. It is an application of Bayes' Theorem: they have a 'prior' likelihood of breast cancer of 8 in 1,000. This is updated by the new information that there has unfortunately been a positive mammogram result.

We see from the information in the question that for every 1,000 women, 77 of them will get a positive mammogram result: 7 who have cancer, and 70 who do not. The likelihood of cancer has been updated from 8 in 1,000 (0.8%) to 7 in 77 (9.1%) after the positive result.

The updated likelihood of cancer, around 9%, comes as a surprise to many people, who would naturally assume that a positive test would mean a much worse prognosis. Even after a positive test, there is still a 91% chance that you do not have cancer. This leads to another debate about the policy proposal. Learning that mammograms give us less information than we thought should make us less inclined to recommend them to women in their forties, given that there are other uses for the money. But a student in the class, who is also a senior doctor at a nearby hospital, chimes in authoritatively:

"The reality is that mammograms are a helpful screening tool. You can't expect them to have high accuracy because the technology isn't good enough: if you want fewer false positives then you can turn down the sensitivity of the test, but you're going to get more false negatives. The problem here isn't with mammograms, it's with our expectations of what a test like that can do."

In bringing her expertise to the discussion, she introduces an important theme: the trade-off between sensitivity and precision. If you have a metal detector, you can turn up the sensitivity and it will start beeping at you whenever it picks up the smallest disturbance, resulting in a lot of false positives. But if you turn down the sensitivity to avoid this problem, it may never beep at all, and you may never find what you are looking for. The same is true of mammograms, and finding the right balance is often a question of policy rather than statistics.

For the next five minutes, the class discusses the doctor's comment. Her expertise contributed to the learning of the class, and created a debate at the intersection of policy and statistics. To some extent, this is Dan's goal: to have the students be responsible for their own learning, and to give them experience at moving from the realm of statistics into the realm of policy.

However, when discussions like this one arise organically from the experience of the students, he has his own trade-off to make between allowing them to continue and returning to the statistical concepts. On this occasion, he felt he allowed the discussion to go on for too long: this is only Class 2, and there will be plenty of time to contemplate precision and sensitivity in future classes. He worried that for those students still struggling to get to grips with the basics of Bayes' Theorem, a technical discussion about mammogram sensitivity may have been a distraction. Some of the more confident students saw it as a helpful debate that extended the topic from theory to practice. There is probably truth in both these assessments: this is one of many examples of the challenge of balancing different needs and expectations across the class.

There is a final part of the mammogram example that brings the intuition home to the students. They did not know it, but Dan was carrying out his own experiment on them when they answered the online question before class about the probability of having breast cancer upon receiving a positive mammogram result. Only half of the students received the prompt above, which started "Eight out of every 1,000 women have breast cancer". The other half received exactly the same information, but in the form of probabilities rather than counts:

"The probability that one of these women has breast cancer is 0.8 percent. If a woman has breast cancer, the probability is 90 percent that she will have a positive mammogram. If she does not have breast cancer, the probability is 7 percent that she will still have a positive mammogram. Imagine a woman who has a positive mammogram. What is the probability that she actually has breast cancer?"

This is a harder problem for the brain to parse than the first one: numbers are more intuitive to us than probabilities, especially when we need to find proportions. Proportions of people make more sense to us than ratios of probabilities. Towards the end of the class, Dan lets the students know about the experiment. This causes laughs and raised eyebrows around the room, and creates a sense of anticipation to see the results.

First, he shows the distribution of the answers from students who were given the question phrased in probabilities. It shows that just over half of the students (56%) got the answer correct at around 9%. However, two in five students (40%) put an answer above 50%, and a sizeable majority of these had answers above 90%. For those students who weren't confident with Bayes' Theorem already, they were likely to fall into the trap of believing their intuition: a positive test result must mean a high probability of cancer.

Figure 2.1: counts are easier to process than probabilities

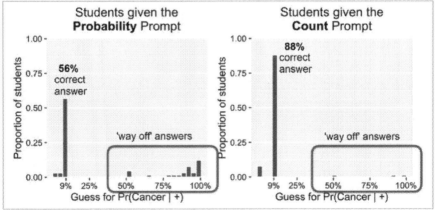

Next, he shows the distribution for the other group, who were given the question phrased in counts. This time, 88% of students got the question right, and 95% had an answer below 10%. Only 5% of students were 'way off', compared to 40% of the other group. The two groups were essentially asked the same question, but the phrasing of that question resulted in massively different error rates.

This is remarkable for two reasons: firstly, our ability to assimilate information depends significantly on the format of the numbers, and secondly, the fact that Bayes' Theorem is most commonly taught in probabilities, not counts. The standard way of teaching it has been to show the formula, $P(A$ given $B) = P(A$ and $B) / P(B)$, and to substitute in probabilities from examples. Expressed in counts instead, the concept is more intuitive to the brain.

EXPLAINING CONCEPTS TO POLICYMAKERS

The knowledge that counts are better for our intuition than probabilities not only helps students to understand concepts more easily, but also to explain them to policymakers more clearly. This is a big theme of the course: in most lectures and problem sets, as well as every final exam for the last fifteen years, there has been at least one question which starts: *"Explain to a policy-maker who is intelligent but not well-versed in statistics…"*

This becomes one of the catchphrases of the class, and by the second half of term it gets a laugh of recognition whenever it comes up. In this class, being able to communicate statistics in clear terms – especially to those who may be in positions of great responsibility – is just as important a skill as understanding the concepts themselves.

It is not a new observation that a good way of testing your understanding of a concept is to try to teach it to someone else. But not all professors make systematic use of the idea: sometimes it will come in the form of an end-of-class presentation, but will not form the basis of the learning itself over the course of the semester. Much of this emphasis on student explanations comes from the positioning of this class as one for policymakers rather than statisticians. But part of it also comes from Dan himself: his own love of teaching and pedagogy, a passion for explaining concepts clearly, and bad memories of his own PhD in which he sat through months of "statistics" classes without feeling like he ever truly understood any of it.

Asked why he puts the students in a position of explaining to policymakers so regularly, Dan responds: "The main reason is that I'm not just trying to maximise learning about statistics, but also to maximise learning of the skills that will be useful to have out there in the world."

This highlights a key objective of the class: to teach students to have a healthy relationship with statistics that will serve them well in their future lives. He continues: "Graduates will rarely be in a situation to solve a problem, write '$x = 5$', put a square box around it, and have that be the end of it. Their life is going to be very different."

Dan is naming the obvious here: the skills we need in the 'real world' are often many steps removed from those we learn in a high school or university math class. This has been a source of frustration for math students in all times and places. But Dan does more than most to bridge this gap, and students appreciate the attempt to do so. In many cases, they will go out into the world and use skills that they practised in his class. Communication is a big part of that, and getting practice in explaining their conclusions with clarity is a valuable part of the course.

As one student writes in their evaluations: "If you can't explain complex topics in simple words, you never really understood them." For those students who have already come across many of the ideas taught in class, the focus on communication helps them to achieve a level of mastery that would have eluded them in previous courses.

The importance of communication is cemented by a task the students must complete in the first problem set. They are asked to use Bayes' Theorem to explore data on the predictive value of HIV testing across different countries. They must then use their results to write a one-paragraph email to a UN development agency with a recommendation on whether a policy of mandatory HIV testing for adults would be appropriate, and in which countries.

The statistical exercise they have done is an important part of answering this question: they find that in countries with low rates of HIV, the test does not predict well enough* to justify a large-scale testing program. But for countries with high rates of HIV, there are ethical concerns to consider too: many would argue that mandatory testing violates the right to privacy. Students are encouraged to weigh up these issues and propose their own compromise.

AVAILABILITY BIAS: BACK TO MARTIN

At the end of the class, Dan tells his students that he knows they are dying to know about Martin, the analytic American who reads the Wall

---

* In the language of Bayes' Theorem, the probability of HIV given a positive test is relatively low.

Street Journal and invests his funds in a well-diversified portfolio. He continues:

"I can tell you now that the majority of you did not get the answer right."

This gets some chuckles in the classroom: students do not mind being told that they are wrong, so long as most of their peers are wrong as well. Dan explains that most students said that Martin was more likely to be a finance professor than a physician. Of course, Martin's traits are closer to those we would naturally associate with a finance professor.

But the key point – and the one that virtually everyone underestimates – is the simple observation that there are many more physicians in the United States than finance professors. A rough estimate is that there are about 1,000,000 physicians, but only 5,000 finance professors. The fact that Martin's traits align more with being a finance professor is easily outweighed by the fact that for every finance professor, there are about 200 physicians.

We can verify this with a simple calculation. There may well be plenty of analytic physicians who read the Wall Street Journal and diversify their portfolios. Suppose that only 1 in 20 physicians share these characteristics, and that *all* finance professors do. Then if we count the number of people who fit the description, we have 5,000 finance professors and 50,000 physicians. The likelihood of Martin being a finance professor, according to Bayes' Theorem, is therefore 5,000 ÷ 55,000, about 9%.

The students' failure to get the answer right comes from the same roots as their failure to accept that Trump had a decent chance of winning the election in November 2016. This is another example of availability bias: the information about Martin leads us to connect him with the profession which most quickly reminds us of those characteristics. Availability bias prevents us from *setting* our prior beliefs appropriately, whereas confirmation bias stops us from *updating* them for new information. In the language of Bayes, availability bias stops us from estimating the prior probability of A, whereas confirmation bias hinders us from updating that probability for the impact of B.

There is another way to think about availability bias. It causes us to mistake 'the probability of A, given B' with 'the probability of B, given A'. In other words, the brain answers the easier question of guessing the likelihood of Martin reading the Wall Street Journal, given that he is a finance professor. This is indeed higher than the likelihood of reading the same paper given that he is a physician. But in the question, his career was event A, and reading the Wall Street Journal (along with other characteristics) was the new information we received as event B. The students who thought he was more likely to be a finance professor unknowingly estimated the probability of 'B given A' instead of 'A given B'.

This may seem like a quirk of the brain: a small imperfection that leads most people to decide against becoming statisticians. But as Dan shows the class in the first pre-class reading, it can have devastating consequences when not understood.

THE TRAGIC CASE OF SALLY CLARK

"It's the same with these deaths. You have to say two unlikely events have happened and together it's very, very, very unlikely."

Professor Roy Meadow, one of the UK's most prominent experts on child protection, had just addressed the courtroom[i]. Based on his testimony, the jury could have little hesitation in its guilty verdict. Sally Clark's baby children had died suspiciously in the same way, a year apart, with no-one around but her on both occasions. This was far too unlikely to leave room for "reasonable doubt".

The rationale was persuasive: cot death happens to roughly 1 in 8,500 unfortunate families like Sally Clark's. The likelihood of having two in the same family was therefore 1 in 73 million: the square of the original probability, or $1/8500 \times 1/8500$. Think of it like two dice: the odds of rolling a 'one' are 1 in 6, so the chances of rolling two 'ones' are 1 in 36.

These are astronomical odds: surely, the combination of events could not have happened by chance. There was no other evidence, but no matter: all evidence has its flaws. The prosecution could point to the one thing more powerful than evidence: mathematics. The math said she was guilty, and the jury agreed.

The only problem was that they were totally wrong. Sally Clark was innocent. The two babies had died tragically, and accidentally. The math was wrong – or, at least, the interpretation was wrong. Professor Meadow was undoubtedly an accomplished paediatrician. However, like most people, he was not an accomplished statistician.

Meadow had made a similar mistake to the one Dan's students made when they said Martin was more likely to be a finance professor than a physician. He had flipped 'A given B' into 'B given A'. He had estimated the probability of the two deaths, *given that* Sally Clark was innocent: 'deaths given innocence'. The relevant figure for the jury to understand was the probability that she was innocent, given the two deaths: 'innocence given deaths'. This form of availability bias, in which the conditionality of the events is reversed by accident, is now known as 'prosecutor's fallacy'.

The flaw in the reasoning may be more apparent if we consider a national lottery, for which the odds of winning are 1 in 10 million. Meadow's logic suggests that only fraudsters can win the lottery. If a person did not cheat, their likelihood of winning was tiny: therefore, they must have cheated to win. It ignores the fact that millions of people had bought a ticket: even though the win was unlikely to happen to any given person, it was likely to happen to *someone*. Sally Clark's case is the opposite of a lottery win. She was the unlucky mother for whom this tragic event happened twice: hugely unlikely to happen to any given person, but likely to happen to someone.

Prosecutor's fallacy was the first, and most famous, error that Meadow made in his analysis for the jury. There were two other egregious mistakes. Firstly, to multiply probabilities together, the two events need to be 'independent' of one another. The occurrence of the first cannot be linked to the occurrence of the second: for example, the events 'it's cold' and 'it's snowing' are not independent, because the first makes second more likely. Meadow assumed that the probability of a second cot death was the same as the first: both at 1 in 8,500. But the science is unclear: there may be a genetic cause of cot death which made the second cot death more likely once the first had happened. If so, the two would not be independent. He should not have multiplied the two probabilities together.

The third error was subtle, but just as catastrophic. Sally Clark was in a low-risk family for cot death: a dual-parent family in which no-one smoked. Overall, cot death has a higher prevalence of 1 in 1,300, but in low-risk families like hers, it is as rare as 1 in 8,500. This was used as evidence *against* her: Meadow argued that the low-risk environment made the alternative hypothesis of murder more likely. He failed to consider that murder is also less likely to happen in low-risk families. The two events – cot death and murder – should have been compared alongside each other, but they were not.

There is no happy ending to the story. Sally Clark spent more than three years in prison for the murder of her two infant children, before her conviction was overturned in 2003. Her life, already in pieces, had been destroyed by the verdict. She spent her time in jail treated as a child murderer. She died a few years after her release at the age of 42, having never recovered her mental health.

The tragedy of her case is a contrast with the joyful and excited atmosphere of Dan's classes. Students are asked to read it before attending the second class on Bayes' Theorem. It helps to give them a sense of purpose and responsibility. Most people do not get the chance to study statistics: it is unreasonable to expect everyone to understand the prosecutor's fallacy. Those who are lucky enough to be in API-209, and especially the law scholars in the room, must take their knowledge out into the world and use it to prevent future cases like that of Sally Clark.

3

RESULTING

Having thought about how to be a Bayesian, and read the Sally Clark case, the students are ready to learn how to put these ideas into practice when decision-making. In advance of the class, as part of the first problem set, he asks the students to complete an online module explaining the basics of 'decision trees'. These trees are set up with square 'decision' nodes whose branches are possible choices, and circle 'outcome' nodes whose branches are the probabilities associated with each outcome. The goal is to be able to 'solve' the tree to work out the right decision.

Dan asks the class for a show of hands. "How many of you had come across decision trees before the module?"

Roughly half of the class puts their hands up. Dan explains to them that about eight years ago, when he first taught this section without the online module, he asked the same question, and got the same answer. "At that moment, I knew I was in trouble", he says. "If I explained the basics of decision trees, half the class would be bored. If I didn't, half the class would be lost. The aim of the pre-class module is to try to get everyone to a level of common understanding, so we can go deeper. This is why you're doing these online modules."

Dan's rationale makes sense here: a little pre-class work can go a long way in bringing everyone closer to the same level. But what is interesting is his tendency to analyse, and then to motivate, his own methods with the students. With lectures, problem sets, readings, office hours and exams to do, the students might be tempted to see the online modules as an unnecessary distraction, especially for the half of them that had seen the content before. Explaining *why* they are doing this extra work is an important part of the contract with the students: it gets their buy-in for the work, and makes them more likely to put in the effort on future pre-class modules.

It also gives the students a heightened sense of responsibility for their own learning. As we will see in Part II, this is crucial to the environment he builds in the class. By bringing the students into his thought process in how he structures their learning, Dan narrows the gap between them and him. The students are thinking one level higher as a result: not just learning statistics, but thinking about *how* they learn it.

LIVING WITHOUT REGRETS

In the online module, one of the questions had asked the students to decide whether, based on the cost and a given probability of loss, they would buy insurance on a new iPhone. As is the case in the real world, the insurance is expensive. There is no right answer, though, since those who are risk-averse may get greater peace of mind from buying it anyway. In class, he asks a follow-up question on Poll Everywhere:

"Given the information in the question, most of you decided you should not buy iPhone insurance. Suppose you decide not to buy insurance on your new iPhone, and 3 months later it breaks down. Will you regret your decision?"

The class splits 63% yes, and 37% no. Dan asks someone who said 'yes' to defend their answer. A woman from Kenya explains that even if the decision was the right one based on the information, the iPhone *did* break, and of course this would lead to a sense of regret. Dan asks for any alternative views.

An American man, John, speaks with confidence: "Well, I guess it depends on what we mean by *regret.*" The class, anticipating a more philosophical discussion than Dan might have expected, starts to chuckle.

Dan pauses and smiles at John, taking advantage of the playful laughter in the room. "Do you always speak like this?" he asks.

This gets the biggest laugh of the semester so far, and it takes a full 15 seconds to subside. Dan continues: "It sounds like you're being interviewed on CNN".

Another 10 seconds of laughter. John is known for his reflective observations, and the students enjoy seeing their confident friend teased.

Stepping away from the subject material, this is a remarkable moment. Many professors might not be able to get away with mocking one of their students. One can imagine an atmosphere of increased tension, rather than laughter, after Dan poked fun at the way John spoke. The key is that Dan has created an atmosphere in which students know that he has their best interests at heart, and one in which they are used to laughing. There are parallels with stand-up comedy: the same joke told by different comedians may go down very differently depending on the atmosphere they have created in the room. We will examine this interaction later in the book.

After a little more discussion, Dan moves on. "How about those of you who said you would *not* regret the decision?"

Shreya, from India, puts her hand up. "At the time, I was making a rational choice", she says. "A change in events later does not make that choice wrong."

Dan probes further: "So, picture this – you're there with your friend, your phone falls and breaks, and you're just going to say 'I made a rational decision' and be fine about it. Are you really that logical?"

The class laughs again. Shreya admits that in the moment, there might be an emotional reaction to not buying the insurance that looks like regret. Dan highlights that the important takeaway is the difference between the feeling of regret and the analysis of the decision:

"You can feel regret if you like: that's an emotion. But was the decision wrong? No. With the information you had, you made the most rational decision. If you had to make it over again, you'll be better off having made that decision."

This leads him to the airport idea for this section: *the quality of the decision is not the same as the quality of the outcome.* For those that play poker, this is a common phenomenon: the top poker players steel themselves not to react emotionally to bad outcomes when their decision was the right one. Annie Duke, who has a World Series of Poker bracelet and over $5 million in winnings to her name, published a book in 2019 called 'Thinking in Bets', in which she included the phenomenon of 'Resulting' – the tendency to equate the quality of a decision with the quality of its outcome. Dan puts the cover on the screen and recommends it to the class, before continuing with the same theme.

"I want this idea to be very ingrained in your minds. I want to pick an example to make this clear. Two friends, A and B, go to a bar. That's how a lot of jokes start, but this is a story. A drove on the way there. At the bar, A drank alcohol and B did not. Who should drive back?"

The class pauses. "It's not a trick question", Dan affirms. Everyone says 'B' in unison.

"So let's suppose A drove, and no accident occurred. Should we conclude this was a good decision?" The class all says 'no' at once.

Again, the quality of the decision is different to the quality of the outcome. Dan draws a two-by-two matrix to hammer the point home. A good decision with a good outcome is 'great'; a bad decision with a good outcome is 'lucky'; a good decision with a bad outcome is 'unlucky', and a bad decision with a bad outcome is 'you deserved it'.

Figure 3.1: the decision-outcome matrix

	Good decision	Bad decision
Good outcome	GREAT!	Lucky
Bad outcome	Unlucky	YOU DESERVED IT!

It is possible that you, the reader, are feeling short-changed. This is supposed to be a course on Advanced Quantitative Methods for the brightest minds Harvard could find. But nothing Dan is saying here is especially advanced. The emphasis on how the 'quality of the decision' and the 'quality of the outcome' are different is hardly a revelation. There are two explanations for the low difficulty level of the material.

The first is that we are still early in the course, and Dan wants to create a positive mindset in which students gain confidence. The concepts *will* get a little tougher.

The second reason is that more fundamentally, statistics is not as hard as some make it out to be. You can make it hard if you want: people do all sorts of doctorates and advanced study in these methods. If you want to prove the mathematical foundations behind all of this, then that can start to get tricky. But the level required to use statistics to interpret the world around you, with assertiveness and authority, is within reach for most people. You do not have to be able to get into Harvard to understand the ideas he is teaching.

STALLING DECISIONS ALREADY MADE

The next exercise for the class is to solve a decision tree. A company called Metropole has two options when installing their new automation technology. Option A has been proven to work, and will cost them $14 million. Option B is a new entry into the market, which claims to be able to install the same technology for $10 million. Metropole estimates that there is a 60% likelihood that Option B's technology will work: if not, they will still be able to pay for Option A, but will face an extra cost of $3 million for the hassle of switching.

Dan gives them a few minutes to conclude that they should pick the cheaper, but riskier option when they install their new automation technology*. Based on the probability of failure and the costs associated, the risk is worth taking. He adds a new branch to the decision tree: if Metropole can acquire a test that will tell them for sure whether the cheaper option works, how much should they be willing to pay for that test? The students recalculate the probabilities and the costs, and work out that the answer is $1.2 million (the test saves them the $3 million cost of switching, 40% of the time).

Dan then takes the same example, but increases the cost of the riskier option. In this second version, Option B still only has a 60% likelihood of working, but costs $15 million: more than Option A, which is certain to work. Clearly, Metropole should now pick Option A, which is both safer and cheaper. He asks the same follow-up question: how much should they be willing to pay for a test which will tell them whether B will work? The class answers in unison: zero. Paying for a test on an option you know you will not use is a waste of money.

Dan observes that the class seems to think that this was an obvious conclusion. "I want to challenge this idea", he says. "We are often in situations in which we seem to be willing to pay for information even though it will not change our decision. I want to define 'pay' as more than just giving money: it could be waiting for more time. We could be

---

* Option B will cost Metropole $10 million 60% of the time, and $17 million 40% of the time, when they switch to Option A. The expected cost is 60% of $10 million + 40% of $17 million = $12.8 million, which is less than the $14 million cost of Option A.

waiting for information to arrive before deciding, even though we know what we're going to do."

He asks the class for examples of when they might have seen this phenomenon in action. A student who was a consultant before coming to Harvard talks about being asked by a government in the Middle East to evaluate every possible option for the location of a new infrastructure project, even though it appeared a decision had already been made as to where the project would be. Dan points out that this is common: governments and companies may be looking for political cover for a decision already made.

"I know there are a lot of consultants in the room, so don't be offended – but often companies ask consultants to tell them to do what they already know they're going to do."

The students laugh in recognition. This is another example of the course concept that statistics is only one part of the puzzle. A policymaker will have to deal with political concerns that are difficult to factor into any quantitative calculation.

It is no coincidence that Dan's class has been able to flourish at the Kennedy School, where the students expect classes to be tailored to future policymakers. Perhaps less obvious is that anyone using statistics in the professional world will also be subject to political challenges. On any decision of consequence, there will be winners and losers. Learning to use statistics to aid decision-making while understanding the *other* moving parts of a system is valuable to anyone – not just those who will end up as policymakers. Statistics is a weaker subject when taught in isolation.

Dan continues with another example, which hits home for most of the students in the class.

"Right around April, you probably received a letter from the Kennedy School saying 'congratulations, you're admitted'. I wonder how many of you were like 'wow, this is my dream', and knew it was what you wanted, but somehow did not inform the Kennedy School until much later, because you were thinking 'I'm still waiting to hear from Kansas State, they might offer more financial aid', or whatever."

The joke at the expense of Kansas State gets a laugh from the class, but the key point is that too often we wait to have all the information before making decisions. We can improve our decisiveness by not waiting for information that will not change a decision.

RISK AVERSION

In the online modules the students completed before class, they had played a mayor of a small town deciding whether to pay a snow removal crew to stay over the weekend given a 40% chance of snowfall. The 'correct' answer is that they should take the risk, and not pay the removal crew. After each of the modules, there is a 'comment' box which invites questions from the students. One of them, Diana, used this box to point out a gap in the analysis. Dan puts her question up on the screen, along with her photo, getting laughs from the class.

Diana is asking: "Suppose I'm extremely risk-averse, and so even if the expected cost of letting the snow crew leave is less, I might want them to stay anyway. How would that be seen in the decision tree?"

Dan tells the class about the idea of 'utility': how happy or sad you would be in a given scenario. Costs translate into utility depending on risk aversion: someone who is risk averse will be happier to settle for a safer but less lucrative option. As is becoming common, he cements the idea with an experiment done on the students themselves. He asks the question on Poll Everywhere:

"You have a choice between two lotteries: the first will give you $20 half the time, and $17 the other half. The second will give you $25 ninety percent of the time, and $0 ten percent of the time. Which would you play?"

The students are all able to calculate that the second is a better lottery on average: you expect to win $22.50, which is more than the $18.50 you expect from the first lottery. This extra $4 is enough for most to compensate the ten percent risk of getting zero: two-thirds of the class pick the riskier lottery, with the remaining third being more risk averse and playing it safe with the first option. Dan continues:

"How about I add a few zeroes to all of these numbers?"

The class chuckles as he draws on his iPad, which is transmitting the question to the screen. He multiplies everything by a million: the new choice is between a fifty-fifty chance of $20 million or $17 million, or a ninety-ten chance of $25 million or zero. The poll numbers on the screen change: now, almost all of the class would take the safety of the first lottery. One student articulates that she expects to make lots of decisions like the one in the first question, but the second would probably be a once-in-a-lifetime opportunity. Dan confirms that this is a common phenomenon: people who are risk-neutral when the stakes are low become risk-averse when the stakes are high.

There's something not quite right, though: five of the class would still choose to take the risk with the millions. He asks them to stand up, which gets a laugh from the class. One British student tries to explain that if he were ever in a position to make that kind of gamble, it would be part of a job in which he was tasked to manage those sums of money regularly – not a once-in-a-lifetime bet. Dan laughs and tells the class: "if he's ever asking you for money to invest, please be careful." The class enjoys the fun poked at the British student: just as before, Dan has created an environment in which a joke at a student's expense goes down well.

ANCHORING AND STATUS QUO BIAS

Dan runs another experiment on the group in this third class. Earlier, when the students arrived, Dan had asked them to fill in the answer to a question on a small piece of paper that has been placed on their desks.

"Please fill in the form and pass it to your left: do *not* Google the answer."

Half the students received a prompt asking whether the population of Turkey was more than 35 million as their first question. The other half received a prompt asking whether the population of Turkey was more than 100 million. All the students received the same second question, asking them to estimate the population of Turkey. Over the course of the lecture itself, one of Dan's course assistants then collates all the answers and calculates the averages for the two groups.

At the end of class, Dan announces that most of the class got the first question right: those who got the first prompt tended to say it was more than 35 million, and those who got the second tended to say it was less than 100 million. The true population is about 80 million. But the important point comes in their answers to the second question. In theory, the number placed in front of them in the first question should not affect their answer to the second. But the students in Group 1, on average, estimated the population of Turkey to be 53 million; the students in Group 2 estimated an average of 73 million. Dan smiles, and tells the class:

"I'm sorry to report that you were victims of anchoring."

He goes on to explain that anchoring is another behavioural phenomenon that leads the brain to make sub-optimal decisions: we rely too much on initial pieces of information which may not be relevant to the final decision. If we are unaware of anchoring, others will be able to use it to their advantage against us. The 'anchoring trap' is used in negotiations to great effect.

Another bias that can be exploited is our tendency to default to the status quo. Dan shows a chart on the left of the screen with organ donation rates in various countries, including the United Kingdom (17%) and Germany (12%), in which those who donate organs must actively opt into the system[ii]. He asks the class to name some similar countries to Germany to highlight the point that any difference is likely to be the result of the system rather than background characteristics of the countries. Students shout out 'Austria', 'France', and 'Belgium', all of which are on the chart about to appear on the right of screen.

When it does, there is an audible 'oooh' of surprise from the class. These are countries with an 'opt-out' system for organ donation: consent is presumed unless otherwise stated. Austria's effective consent rate is 99.98%, France 99.91%, and Belgium 98%: almost no-one chooses to opt out. With organ donation, we are much more likely to choose the default option, whatever that option is. People might have been expecting an increase, but the size of the gap between the opt-ins and the opt-outs comes as a shock.

Dan ends the class with a final observation:

"You can use this insight to do great good in the world; you could also use it to do not so good. My hope is that you will choose the former."

Once again, the day's learning is motivated by the impact his students will one day have in the world. These behavioural ideas are closer to psychology than statistics, but without understanding them, we risk drawing false conclusions from data. We miss out on opportunities to use the way the mind works to our advantage, and we let others take advantage of us.

4

THE CURIOUS CASE OF MEXICAN PENSIONS

The learning from the first three classes culminates in an immersive case study of pensions in Mexico: an application of skills that serves as the "end of the beginning" of Dan's course. When students graduate, about twenty months later, many still remember this fourth class as one of the highlights of their time at the Kennedy School.

The foundation for the class is built with the work that students must complete for the first of the problem sets. Throughout the course, problem sets will not just reinforce ideas covered in class, like the example of Bayes' Theorem and HIV testing, but also prepare and motivate upcoming learning. Assignments are entirely case-based: there are none of the repetitive number-crunching exercises that many associate with 'math homework'.

The students are asked to read a case study preparation document entitled 'Providing Pensions for the Poor: Targeting Cash Transfers for the Elderly in Mexico'. The document explains that in 2006, Mexico had a serious problem with high poverty among adults aged 65 and over. There was a big rural-urban divide, with 36% of seniors in urban areas in poverty, compared to 55% in rural areas. This came despite a successful cash transfer program named *Oportunidades* ('Opportunities'), which helped to reduce poverty overall but had little positive impact on the elderly. The new President, Felipe Calderon, was seeking to introduce a new program which would more accurately target those elderly people in need.

They would also need to keep in mind that they had a yearly budget of 8.5 billion pesos (about $800 million) for the program, and were seeking to give eligible seniors 500 pesos per month. This meant that they had the funds to cover about 1.4 million people with the new program.

There were three options proposed:

- **Option 1:** All individuals 70 years and older living in a household covered by *Oportunidades* would be eligible for the program.

- **Option 2:** All individuals 70 years and older living in rural areas (those of 2,500 or fewer inhabitants) would be eligible for the program.

- **Option 3:** All individuals 70 years and older living in areas with weak access to education, housing, water and sanitation (as measured by a government index) would be eligible for the program.

The case ends with the students' mission:

"The Minister of Social Development has asked you, the Director General of Planning, to evaluate the three options and recommend one. She expects you to base your recommendations on a broad set of criteria including targeting efficiency, financial feasibility, logistics, and politics. Your deadline is tomorrow."

LEAKAGE AND UNDERCOVERAGE

On beginning the mission, students immediately feel pushed out of their comfort zone. No longer are they working with toy examples: this is a real case, and they must make a real recommendation based on the same information that policymakers in Mexico had. Their first task is to make some sense of the numbers in the appendix of the case document.

The appendix explains that there are two main errors that can be made when targeting a poverty program. Firstly, you can give the money to someone who is not poor: this is called 'leakage', as the money is 'leaking' out of the intended pool of recipients. Secondly, you can fail to give money to someone who *is* poor: this is called 'undercoverage'.

There is a natural trade-off between the two: you can achieve zero leakage by not paying out any money at all, but then you also have 100% undercoverage, as no poor people get money. You can achieve zero undercoverage by giving money to everyone in the population, but this means huge leakage as all the rich people are getting money as well. Since the government does not have the ability to target the poor perfectly, they will have to find some optimal trade-off between leakage and undercoverage.

These concepts can be expressed in the language of probability and Bayes' Theorem. The leakage rate is the proportion of those enrolled who are *not* poor, and the undercoverage rate is the proportion of those who *are* poor who are not enrolled[*]. Students are given the numbers required to do these calculations. Across the country, there are about 4.6 million people over 70, of which about 1.9 million are poor.

Figure 4.1: number of individuals aged 70+ covered by the three options

	Total	Poor
Number of individuals aged 70+	**4,592,726**	**1,872,313**
Option 1: live in a household covered by *Oportunidades*	859,299	637,897
Option 2: live in a rural area of less than 2,500 people	1,413,182	801,537
Option 3: live in an area with weak access to essentials	3,040,844	1,047,934

Based on this table, the students can calculate the leakage and undercoverage rates for each of the programs. The leakage rate in the first program is relatively low: it gives money to about 859,000 elderly people, of which about 638,000 are poor. That means that there are 221,000 people receiving money who are *not* poor, which is 26% of all recipients. However, the undercoverage ratio is high: there are 1.9 million poor people over 70 in the country, and only 638,000 are being

---

[*] In probability notation: leakage = P(not poor | enrolled), and undercoverage = P (not enrolled | poor).

given money. That means that 1.2 million of them are *not* given money: an undercoverage ratio of 66%.

The completed table, which the students are required to report in the problem set, for all three options is shown below. They find that the three options come in increasing order of leakage, and reducing order of undercoverage:

Figure 4.2: leakage and undercoverage ratios for each of the three options

Option	Leakage	Undercoverage
Option 1: live in a household covered by *Oportunidades*	26%	66%
Option 2: live in a rural area of less than 2,500 people	43%	57%
Option 3: live in an area with weak access to essentials	66%	44%

In the usual math homework example, this would be the end of the question: having shown that they can calculate leakage and undercoverage rates, the students would be able to breathe a sigh of relief and move on. Not so for Dan's problem sets: the students are then asked to lay out the advantages and disadvantages of each approach, then write a one-paragraph email to the Minister of Social Development to make their recommendation. As usual, they are told to assume that the minister is intelligent, but not well-versed in statistics.

Dan believes that the key advantage of this class is that "it presents the world as it's coming to you". In simulating the situation that students may one day find themselves in, he blurs the distinction between theory and reality. Once the problem set has been completed, which for the average student takes eight to ten hours, they are asked to come to class prepared with their analysis and opinions.

GETTING THE STUDENTS EXCITED

At the start of the lecture, Dan tells the students that this will be a change from the first three classes, and that he expects more hands to go up than in a typical class. To manage this, he establishes ground rules. The first is that students must build on each other's comments: if not, it becomes a free-for-all.

He clarifies: "This does not mean just saying 'building on what X said', and then saying something totally unrelated, but substantive building on each other's comments. That means that if someone is talking and another has their hand up already, I know they're not going to build on them."

The class laughs: at Harvard, where students are not known for lacking confidence in their opinions, it is a common tactic to pretend to build on what someone else said while ignoring it completely. Dan is recognising here that the norms he established in the first class may have to be bolstered because of the unstructured debate he wants the class to have. The less he is forced to intervene during class to bring unrelated comments together, the more the discussion will be able to flow.

He continues: "The second rule is that I'd like you to engage as if you were actually working for the Ministry, and defending the choice you made. I want this to be an intense discussion, where you defend your opinion passionately. I want that energy level in the classroom."

This discussion is important to the class, and Dan knows that if it goes well, it will be memorable for the students. He uses emotive words like 'intense', 'passionate' and 'energy' to ensure everyone is focused. This kind of motivation should be used sparingly: if Dan were to introduce *every* class with that level of intensity, it would quickly lose its effect. For this one, though, he wants the students on the edge of their seats.

HOW WOULD WE TWEET THIS?

Dan announces that he would like someone who has not yet participated in class to tell him how they calculated the undercoverage ratio for Option 1. He already knows who these people are, thanks to an app called Teachly that we will see more of in Part II. The goal is not to name and shame those who have kept quiet, but rather to ensure everyone is included, and not to call on the same people too often. Given that this is a question the whole class has answered in their problem sets, there should be plenty of hands, even among those who would otherwise be reluctant to participate. This also establishes a norm: no-one is expendable in this class. Everyone is expected to contribute, and to support the learning of everyone else.

An Australian student, Chris, makes his first comment of the course, and correctly describes how he arrived at the 66% undercoverage ratio for Option 1.

Dan asks him if he can tell him how he got to the leakage ratio as well, and Chris does so. "We look at the people who are eligible, and how many of those are non-poor. That ratio is 225,000 divided by 859,000, which is about 26%."

"That's great", Dan continues. "You said it in words that your grandma could understand. Now how would you express that ratio in the language of probability and Bayes?"

Chris pauses. Dan looks at the class, and starts to write on the board, suggesting an answer himself. "Is it... the probability of being eligible given that they're non-poor?"

The class shouts out "No!"

"The other way round?" he suggests.

"Yes!" the class replies, as if in a pantomime.

"Are those two things the same?"

"No!", says the class, recalling the importance of not flipping the conditionality by accident, as seen with Martin the investor and the Sally Clark case.

Dan corrects the board from P(*eligible | not poor*) to P(*not poor | eligible*). He is trying to make sure the students feel comfortable changing between the mathematical language of probability, and the more informal language that one would use in a conversation.

He continues: "Now, let's express the whole thing in a language everyone can understand. How would we tweet this?"

Most of the class is on Twitter and will relate to the challenge of making complex ideas simple in the space of 280 characters. An American student tries to recreate the definition of leakage as best he can, and says: "of all the people eligible for the program, 26% of them are not actually poor; these are resources that are wasted by people that shouldn't receive the benefit."

This is correct, but sounds a bit clunky for Twitter. Dan encourages him to remember the lessons from the second class about numbers being easier for the brain to process than percentages. They end up agreeing on "1 in every 4 people in the program does not deserve to be there." There is a murmur of appreciation in the class at how they have distilled a statistical idea into a sentence so crisp and clear that no statistical background is required to understand it. Dan invites a Japanese student to try something similar for the undercoverage ratio of 66%. He obliges: "2 out of every 3 poor people are not receiving the program."

Dan then gets the class to think about the trade-off. A French student outlines that you can either be strict about who gets the program, or you can be generous and give it to more people. The first leads to low leakage but high undercoverage; the second leads to low undercoverage and high leakage. Dan connects the idea with trade-offs the class has already seen, and those that are yet to come:

"This is super-important: you have a natural trade-off between leakage and undercoverage. This applies to many aspects of policy. So when we were doing the mammogram example two classes ago, it was between false positives and false negatives."

As he highlights, the size of the program is analogous to the sensitivity of the mammogram test. Increasing sensitivity (size) will lead to more false positives (non-poor people getting money), but fewer false negatives (poor people missing out). This trade-off is a recurring idea in statistics, and he wants to be cemented in students' heads before they see it again next week amid other concepts that might be new to them.

WHICH OPTION IS BEST?

Moving on, Dan now wants the class to defend their decisions.

"I'm going to let Option 1 supporters – about 40% of the people in this room – tell me everything they liked. For those of you that did not pick Option 1: for the next three minutes, breathe. And prepare."

Figure 4.3: student choices for the three options

People covered by the program	% students
Option 1: live in a household covered by *Oportunidades*	41%
Option 2: live in a rural area of less than 2,500 people	38%
Option 3: live in an area with weak access to essentials	21%
TOTAL	100%

An Indian student, Rajiv, says that he looked at the ratio between leakage and undercoverage, and selected Option 1 based on that ratio. He also says that the program can build upon on the existing *Oportunidades* scheme: the government already knows who to give money to. An American student chimes in to agree: "I really liked that it was the smallest implementation of the options: for a brand new scheme it's a good base to expand from."

Dan nods: "You're starting small – you like that. Pedro, do you agree?"

Pedro, who had tentatively raised his hand to make his first comment of the course, does agree: "Yes, it shows that they're looking to use resources efficiently. They should also have the political support of the President given the low leakage, which is something he will be happy with: they won't overspend as a result."

Ana, from Spain, agrees too, and adds: "It's targeting the same people as Oportunidades. We can expand the programs together: there will be synergies."

An American student, Oliver, also agrees, and also points out that since older people in Mexico tend to live with family members, targeting based on the same households as *Oportunidades* makes sense.

There is a lot of positivity in the room for Option 1. However, Ndidi, a student from Nigeria, has her doubts. "I'm wondering how this is going to get us votes for the next election. It doesn't seem so different from Oportunidades. I'd want something with a much broader reach."

Dan smiles: the political angle is an important one for the class to be aware of. Gabriela, from Chile, builds on Ndidi's comment. "Also, how about the political cost of *not* spending the budget? They said the program cost is 500 pesos per month, so we're spending 5.2 billion, which is way under the 8.5 billion we've been given."

Dan interrupts: "What's wrong with underspending? In the private sector, you get rewarded for that, right?"

Gabriela responds instantly: "Not in my country, in the public sector!" The class laughs, and she continues: "If you have an approved budget, it is not good to underspend."

Dan turns to Francisco, a Mexican student with some experience in this area. "Francisco, I know you worked in the Ministry of Finance. Tell us a little about this. If the Ministry authorises a budget for a program, and they don't spend it, what happens?"

"Congress approves the budget yearly", says Francisco, "so you'll probably reduce the money available for the program next year."

"Is that good?" asks Dan to the rest of the class.

"No!" say many of them in unison, though there are a few furrowed brows, especially from those students with private-sector backgrounds, at the idea that going under-budget is a bad thing. In interviews, Dan mentions this moment as an important learning point that comes up every year: "People with government backgrounds always think that underspending the budget is terrible, while those from the private sector are fine with it." Understanding how different people *think* is a key skill in the program, and in life.

Juan, from Spain, has his hand up. "My concern with Option 1", he says, "is that we're giving extra benefits to the same people who already get money from Oportunidades. We've been talking about poverty as if it's a binary thing, but everyone's on a spectrum. It's the ones that the existing system is missing that we need to help the most."

Dan nods. "Let's examine this a bit more. This program is going to where Oportunidades is going. Where is it being implemented?"

Steven, a British student, points out that *Oportunidades* is good at targeting poor people in urban areas, and bad at getting the ones in rural areas. Dan agrees, and comes back to Steven's point from earlier that older people tend to live with family members.

"What about the ones who live without their families?" he asks. "It's true that lots of them live with children, but those who don't may be some of the most vulnerable. That might worry us if our goal is to target the very poor."

Virat, from India, has an objection to Option 1 linked to Dan's point. "Lots of these social security schemes are formed based on a moral argument. I would want to minimise undercoverage: from a humanitarian perspective, you want to cover as many poor people as you can. Option 1 has the worst undercoverage ratio of the three."

Dan replies: "Okay, so you don't like Option 1 because for you, undercoverage is more important than leakage. Let's have a show of hands. Everyone, in your roles working for the Mexican government: if you had to prioritise one over the other, which would you choose?"

The class is divided roughly 50-50: those wanting to minimise leakage have a close eye on the Ministry of Finance and the cost of the program, while those for undercoverage, like Virat, are arguing from a humanitarian perspective.

"So what we're seeing here", observes Dan, "is that we're split. Why is that? There's not a technically correct answer to the question of which you should minimise. Values drive a lot of what you do: you can't say statistically that you should do one thing or the other."

He continues: "You *can* start to look at the trade-offs. Moving from Option 1 to Option 2 increases leakage by 18%: perhaps the 9% improvement in undercoverage is enough to compensate it, perhaps not. Which one you value the most is not going to be in the numbers, and that's okay. Policy is not just driven by technical criteria."

Moving on, Dan asks the students who picked Option 2 to explain why they did so. In this scenario, the government would target all villages with a population below 2,500, which is the government's definition of 'rural'. A student points out that the cost of the program, 8.5 billion pesos, is in line with what was budgeted.

Dan agrees: "Okay – we know from the previous discussion that underspending is not good, and here you're spending all the money. What else?"

A couple of students point out that Option 2 starts to address the issue of the rural-urban divide by specifically targeting small villages, and covers as many poor people as it can while factoring in budget constraints. Just like it did with Option 1, momentum starts to gather in the class behind Option 2: it's on budget, and students suggest that it seems to be a good compromise between leakage and undercoverage.

Dan has a warning for the class. "How many know that story about Goldilocks?" Some of the class indicate they do, but there are a few puzzled faces. "I'm not sure how universal it is. Anyway, let me just say that it's often we see a table like this and we're attracted to the middle option. It's not too leakage-y, not too undercoverage-y."

The class starts to chuckle in recognition. Dan continues: "You have to be careful. We tend to fall prey to liking the middle option, which often means that someone might present you with three options that are engineered such that *their* preferred option is the middle one. Apparently, this happened in Vietnam: President Johnson was presented with three options that looked something like 'blow up the world', 'continue the course of action', or 'retreat in defeat'. They made a bad option look like the sensible choice by putting it in the middle. Okay, so the Vietnam War was much more complicated than this, but you get the picture."

A Colombian student, who is a fan of behavioural economics, points out that this is called the 'decoy effect'. Just like anchoring and status quo bias from the last class, these are psychological traps that can be used against us if we are unaware of them.

Charlotte, from Taiwan, has some scepticism about giving money out purely because seniors live in a village with under 2,500 inhabitants. "Not all of these will be poor villages", she says. Prina, from India, agrees, and points out that there is a political cost to leakage too: in her experience, the poor do not like seeing the non-poor getting benefits ahead of them.

Dan confirms that this is a risk, and emphasises Juan's point from earlier in the class. So far, they have been discussing poverty as if being poor is a binary characteristic, but there is a spectrum. Leakage may not be such a bad thing if the money is going to people who are poor but just above the cut-off. The problem that Charlotte raises is the potential for "gross leakage" – money handed out to those who are already well-off.

Zhao, from China, has experience in implementing a similar kind of benefits system in his own country. He points out that although the total cost of the disbursements might be just within budget, it will be challenging to implement a program focusing on rural villages, many of which will be geographically isolated. This will lead to high transport and administration costs: another disadvantage of Option 2.

Just like Option 1, which focused on households already on the *Oportunidades* scheme, Option 2 initially appears an excellent choice but flaws are revealed as the students carry out their discussion. Dan moves on to Option 3, which would hand out money to all those considered poor according to the government's poverty index. This looks not just at income, but also at access to education, healthcare, water and sanitation.

Mariana, from Colombia, makes her first contribution of the course. "I liked how they're targeting based on a multi-dimensional poverty index. It seems much better to do it that way than to just hand out the money based only on the size of the village."

Cristina, from Peru, agrees: "This is great, because the targeting criteria are more robust. So even though it has higher leakage than the others, it's likely that even if you're not so poor that you're in the target population, you're still quite poor."

By this point in the class, with Juan's earlier intervention having been magnified by Dan, students are steering away from thinking about poverty in binary terms, even if the necessity of targeting leads it to be defined that way.

Steven, the British student, also mentions that the deprivation index sets it aside from the *Oportunidades* program. "That's politically appealing", he says, "because it's different to the past regime, which they've spent a lot of time criticising."

Dan, who has been tracking the advantages and disadvantages of each method on the blackboard, smiles. "You'll see that all three of them have 'political support' listed as an advantage. I don't want to be wishy-washy here, but they do all have advantages that you'll have to weigh up against each other." He continues: "There is one big problem with Option 3 that people have hinted at, but not mentioned yet."

Mateo, from Venezuela, answers immediately. "It's not fiscally sustainable."

Dan smiles. "Those are big words. What does that mean?"

Mateo replies: "You don't have enough resources."

In big chalk letters, Dan writes "NOT ENOUGH $$$" next to Option 3 on the board. He points out that the total budget is $8.5 billion, which at 500 pesos per month means they can only serve 1.4 million people. They would need to serve about 3 million under Option 3. He asks the class: "Is this a big problem or not?"

Ling, an American student, answers: "I think this is a huge problem. The non-poor may have the means to get disbursements quicker than the poor. By the time we've gone through them, we will have run out of cash."

Dan nods, and continues with an analysis of students' answers from the first problem set. "There's a sizeable part of the class that picked Option 3, and they fall into two categories. The first group just ignored the finances. A big lesson from this case, even if it's in a Statistics problem set, is that you can't ignore how much money you have. The second group said 'oh yeah, we went over budget, but we'll do something else about that." The class laughs.

Dan concludes: "That's great spirit, but not how the world works, most of the time."

THE DEBRIEF

On the blackboard are all the pros and cons that the students identified. A central takeaway for the students is that there is no perfect option: all of the options available had sizeable flaws. Choosing between flawed options will be the norm for those students who later find themselves with decision-making responsibilities in government.

Dan believes that it makes sense to prepare them for that now. "This discussion was very rich", he says, "and exactly the discussion I wanted to have. A big lesson is that there was no option that was the best option: there are real trade-offs. They come not just within the technical criteria – leakage and undercoverage – but also between the technical, political and administrative challenges."

Dan changes tack: "I also want to tell you that this case was written by a former student of this class. He's working with the government in Mexico, and helped to advise them on the choice that you all just debated." He pauses for effect.

"I'm happy to tell you that we can go live to him now."

There is an intake of breath from the students. "Antonio, are you there?"

The screen at the front turns to Antonio, who is there by video link from his office in Mexico City. The class lets out an audible "wow": this is one of Dan's favourite moments from the program, and one of the most memorable for the students. Antonio has been there in the background the whole time, and has listened carefully to the discussion.

The class is on edge to find out which of the options was chosen in real life. After introducing himself, Antonio reveals: "When the law was sent by the President to Congress, Option 2 was the one chosen. Option 1 was given lots of consideration; Option 3 was a decoy put there by Dan and me. We wanted to make the point about the importance of finance and the budget."

The 40% of the class who picked Option 2 are feeling satisfied with themselves. Antonio continues: "The reason Option 2 was chosen was that the point on underspending was crucial. It's a political thing too: you'd be sitting on money that could be benefiting poor people. The President had a very slim majority when he was elected, and this program had been created by the opposition in Congress by taking little bits of budget from lots of other places. Turning round and spending less would have been a slap in the face to Congress, and he didn't want that. The next year, the program was expanded to villages of up to 5,000 people; the year after that, up to 10,000; now, it's nearly universal."

This issue of the politics around underspending had been highlighted by Ndidi and Gabriela in class. Antonio pauses, then tells the class that the quality of the discussion was high, and jokes that most of them would be ready to work in the Mexican government.

He goes on: "The other thing that had a big weight in the decision was that Oportunidades had been a successful program after decades of failures. Part of that success was that it was focused. As a result, people did not want to overcomplicate this new program with lots of complicated requirements. Leakage would have been to households who weren't very rich, but undercoverage could have hit some of the most vulnerable people."

This confirms the points made by Charlotte and Steven in class: in the trade-offs between Option 1 and Option 2, which were both within budget, leakage was 'better' than undercoverage because the money would mostly have 'leaked' to people just above the poverty line. Antonio pauses again, and mentions that the mayor of Mexico City at the time, Andrés Manuel Lopez Obrador, took a lot of credit for the success. Lopez Obrador, colloquially known as 'AMLO', is now the President of Mexico, having won the national election in 2018.

Antonio concludes: "Finally, another important reason was that by targeting at the village level, once a village was in the program, everyone got the benefit. You avoid the ethical dilemma of 'why is my neighbour getting it and not me?' There's not much difference between someone just above the poverty line, and someone just below."

Antonio takes questions from the class, but there's not enough time for everyone. It's noticeable how many students have their hands up: their immersion in the case, first in the problem set and now during class, has led to a high level of engagement. Students are interested in the political dynamics that Antonio has highlighted, as most had focused more on the technical aspects of the decision in their problem set work. One student, from Argentina, wonders why it was a binary decision between two options, rather than a whole range of possible ideas that the President and Congress should be able to choose from.

Antonio answers that time was a big factor: there were only ten days for the President to sign it into law after Congress had approved the budget. They needed to simplify the decision to ensure it was made in time, while restricting their attention to programs that could be implemented quickly. In Congress, lawmakers wanted quick wins: they would ask how many people could be reached in the first year. Option 1 was the obvious choice for many people, especially among the technical advisors like Antonio. They had been forced by policymakers to think more about the political consequences, and came up with Option 2 in response.

Time is up for the class, but several students stay behind to continue talking to Antonio over the video link. For many, this class has been a defining moment in their time at the Kennedy School. It is the class in which they realised that understanding statistics is a great help in making decisions, but does not get them all the way. If they want to use their statistical conclusions to make change, they will need to understand the politics as well.

Having spent some time in Dan's classroom, our goal for the next part of the book is to try to uncover some of the 'magic' behind it. Most of what happens in a classroom is invisible to the participants, and the goal is to bring these invisible dynamics to the surface. We will spend some time thinking about what happens when we learn, and how teachers can bring about the kind of deep learning that sticks with us for years after a class is over.

We will aim to develop some tools to analyse his teaching. Teaching is a form of leadership, the study of which considers how members of a system can confront challenges by mobilising the resources within it. We will consider the invisible building blocks of the learning environment, as well as the various competing purposes that students and teachers are dealing with in class, and think about the tools that Dan has at his disposal to maximise learning. Once we have done so, we will return to the classroom in Part III to apply these ideas to the next part of his class.

PART II: THE INVISIBILITY OF LEARNING

5

IF LEARNING WERE VISIBLE

Imagine a world in which whenever you learned something, its value to you was immediately visible above your head and in your bank account. For example, if you are destined to end up as a successful engineer, then perhaps a big $10,000 appears in your account when you learn Pythagoras's theorem for the first time. The number appears with some kind of 'ka-ching' sound, like in a video game where completing tasks earns points.

In this world, the competition for teachers would be fierce. They would earn even more than bankers, as great teaching would immediately be linked to high rewards. It would be a no-brainer to pay large sums of money to the best teachers, because you would know that previous students made instant profits from attending their classes and learning key concepts. Teachers would do their best to maximise the value of the learning they were creating. As soon as a concept was understood, the money would appear and the student would know about it. They would also be able to track how much they 'made' by learning from each teacher. In turn, the teacher would be able to track which explanations made the most money, and fine-tune accordingly.

It is a strange thought that the only difference between this world and ours is the lack of visibility. It is difficult to measure the returns to education, and almost impossible to measure the value of a specific teacher, let alone a specific class or explanation. We often fail to pinpoint *when* we learned a new idea, let alone put a value on it based on our unknown future. Some concepts illuminate themselves suddenly in 'aha' moments, while others creep up on us slowly.

Teachers earn far less than bankers in this world because learning is invisible. As a result, they are far removed from the value they create. The elementary-school teacher who taught design to Steve Jobs probably never saw any of the rewards from the class that year, even though their actions led to the creation of the iPhone. In contrast, a banker who helps two companies to merge can proudly show an Excel file with the amount of value created, and the hefty fee they were able to take for doing it. Banking is a discipline in which value creation is rewarded immediately.

Although it is very hard to pinpoint the value of a specific teacher, there is plenty of evidence that the 'teacher effect' – the variation in student outcomes that can be explained by the quality of their teachers – is large. One might think that it is the *school* that is the most important factor in explaining how well children do. In fact, the luck of the draw in which *teacher* you happen to get within your school is more important. Teachers matter.

Imagine two identical students, who go to different schools: one rich, one poor. Despite the lack of resources in the poor school, if the two students get teachers of similar quality, the difference in outcomes will be low. However, if the two students both go to the same school, and one gets a better teacher than the other, the difference in outcomes will be much higher. The American economist Eric Hanushek has studied this phenomenon closely. He writes in a 2010 paper[iii]:

> *"Average gains in learning across classrooms, even classrooms within the same school, are very different. Some teachers year after year produce bigger gains in student learning than other teachers. The magnitude of the difference is truly large, with some teachers producing 1½ years of gain in achievement in an academic year while others with equivalent students produce only ½ year of gain. In other words, two students starting at the same level of achievement can know vastly different amounts at the end of a single year due solely to the teacher to which they are assigned. If a bad year is compounded by other bad years, it may not be possible for the student to recover."*

The same paper estimates the value of a 'good' teacher[*] as $20,000 *each* in a class of 20, for each year of teaching. If he is right, then a fair salary for a good teacher would be at least $400,000 given the value they create. In our fantasy world in which we earn while we learn, these numbers would not be unreasonable. However, even though research can put an estimate on the huge value of teachers, the invisibility of learning means that it has not yet been possible to

---

[*] Here, 'good' is defined as one standard deviation above average. Put another way, a 'good' teacher is one who is better than five out of six teachers.

measure *what* good teachers do to create so much value. As Hanushek writes:

> *"Literally hundreds of research studies have focused on the importance of teachers for student achievement. Two key findings emerge. First, teachers are very important; no other measured aspect of schools is nearly as important in determining student achievement. Second, it has not been possible to identify any specific characteristics of teachers that are reliably related to student outcomes."*

In this section, we will explore some of the challenges associated with the invisibility of learning. The goal is to shed light on the invisible dynamics in the classroom, and to understand what contributes to the learning environment in Dan's class. The first key idea is that when you have a challenge for which measuring progress is hard, the ability to adapt is just as important as the ability to plan.

LEARNING AS AN ADAPTIVE CHALLENGE

If you were to walk out of Dan's classroom at the end of an API-209 lecture, go past the windows overlooking the Kennedy School courtyard, down the stairs to the lounge, and walk across the forum where the world leaders speak in the evenings, you would find yourself in a class on 'Adaptive Leadership', taught by Professor Ron Heifetz. He has been teaching students how to change the world for nearly forty years, and his book 'Leadership Without Easy Answers', published in 1994, is a classic in the field. Many of his students are not fresh-eyed graduates, but mid-career CEOs and government leaders who want to be more effective.

No-one ever comes out of the course with the same idea of 'leadership' they had when they went in. Most assume that they will be taught how to be bold and assertive in their decision-making, or how to be more charismatic. By the end of the class, they realise that exercising leadership is more about being a good listener. Mobilising change in a system often needs little more than for people to feel heard.

One of the first things that students learn in Heifetz' class is to distinguish between *technical* and *adaptive* challenges[iv]. Technical challenges are those for which progress is visible and foreseeable. If you are working with NASA, building a rocket would fall into this category. It may be hard, but you can make a plan and monitor it every step of the way. If someone asks, "when were the boosters attached?", you can check the log and give an exact answer. Building a rocket takes time and effort, but you can see your progress throughout.

Adaptive challenges are those for which progress is *invisible*, or at least unforeseeable. No-one knows how to do them, which means they are subject to considerable uncertainty. If you are attempting to get over a painful break-up, you have an adaptive challenge. There is no formula you can use to tell you the exact steps, and you will have good days and bad days. Maybe the day will come when you suddenly feel like a new person; perhaps it will be a more gradual process in which you cannot pinpoint the moment of success. You may have made substantial progress, but it will be invisible and unmeasurable.

A single challenge may have technical and adaptive components. Imagine you are offered an exciting new job on the other side of the world, but you would have to leave your current job and move away from your family to take it. Part of the challenge here is technical: you need to hand in your notice, buy the plane tickets, and sort out accommodation. It will be visible and obvious when these tasks are completed. The more significant part of the challenge is adaptive: you will have to confront the losses that come with living away from family, joining a new team, and getting used to the new environment. The adaptive components are the 'scary' part of the challenge: the invisible leap into the unknown.

Adaptiveness is closely tied to the invisibility of progress. When progress is visible, we can make a plan, and tick boxes when we see the relevant task is done. We can be mechanical about it: having made the plan, we do not expect to deviate from it. When progress is *invisible,* we cannot plan in the same way. If it turns out we were going in the wrong direction, we will have to change course.

Harvard's Matt Andrews, who teaches the 'Getting Things Done' class for policymakers at the Kennedy School, gives the example of the Lewis and Clark explorers in 1803. They were asked by President Thomas Jefferson to lead an expedition travelling west from St. Louis, then the western frontier of the United States. With the map invisible to them, they had to be adaptive in their approach, finding paths around mountains and over rivers, and at each stage bargaining with local indigenous people to help them understand the area and progress safely. It was impossible for them to know how far they had progressed relative to the whole landmass of the United States, but they eventually arrived on the west coast three years later.

Andrews compares the adaptive challenge of Lewis and Clark to the equivalent problem in 2020. If you want to go from St. Louis to the west coast now, all you need is a plan. Google Maps will tell you the quickest path to take, and you can avoid having to adapt by planning fuel stops and lunch breaks. There are always things that can go wrong if you are unlucky: you might be confronted by a maniac driving the wrong way down the road. But this is largely a technical challenge, and your progress will be visible. Lewis and Clark's expedition, combined with technological advances, have turned an adaptive challenge into a technical one.

It is a central argument of this book that learning is an adaptive challenge. We do not know in advance how long it will take to fully understand a new concept, and often we fumble around in the dark before it finally 'clicks' and the light is switched on. Sometimes we never experience that 'click' moment, but our understanding still improves over time. We find ourselves looking back after a class has finished, able to use concepts with dexterity, but unable to pinpoint the moment of illumination. In this case, the light has been slowly increased as if from darkness to dawn.

Emily, an American student interested in Dan's pedagogy, talks to me about this moment of illumination. "I really enjoy seeing people have that 'aha!' moment of understanding. Dan is incredibly good at guiding students to that moment, and helping concepts resonate and stick in an intuitive and practical way."

The key word here is 'guiding'. Since we are not in the fantasy land in which learning is visible and quantifiable, we cannot clearly see our progress. We are more like Lewis and Clark than a modern-day route planner. The Ancient Greek mathematician Euclid, when asked by his king for a shortcut to avoid having to read his book, told him that "there is no royal road to geometry". Today, there is no Google Maps for learning. Our progress will often be invisible and hard to measure, and we will need all the help we can get.

MEMORIZATION, UNDERSTANDING, AND "CHUNKING"

Charles Dickens' classic novel *Hard Times*[v] starts in a "plain, bare, monotonous vault" of a classroom. The man at the front, instructing the schoolmaster, has an unambiguous approach to education:

> *"Now, what I want is, Facts. Teach these boys and girls nothing but Facts. Facts alone are what is wanted in life. Plant nothing else, and root out everything else. You can only form the minds of reasoning animals upon Facts: nothing else will ever be of service to them... Stick to Facts, Sir!"*

The children, as described by Dickens, were "little vessels then and there arranged in order, ready to have imperial gallons of facts poured into them until they were full to the brim."

Education theory has come some way since the schoolmasters of nineteenth-century Britain. The classroom, in which students listen obediently and learn to parrot back facts, is the opposite of Dan's, in which certainty is an illusion and students lead much of the discussion. This kind of old-fashioned classroom is a bad model for learning, for reasons we shall discuss in the next chapter, and which may be obvious to you already.

However, we must not be too quick to dismiss memorization of facts as a learning tool. Facts *are* important for learning. The cognitive neuroscientist Daniel Willingham wrote a book in 2009 entitled 'Why

Don't Students Like School?'[vi] in which he points out that the science supports the importance of memorization:

> *"There is no doubt that having students memorize lists of dry facts is not enriching. It is also true (though less often appreciated) that trying to teach students skills... in the absence of factual knowledge is impossible. Research from cognitive science has shown that the sorts of skills that teachers want for students – such as the ability to analyze and to think critically – require extensive factual knowledge... Factual knowledge must precede skill."*

One reason for this can be found in the writings of George Miller, one of the founders of cognitive psychology, in 1956. He wrote a paper entitled 'The Magical Number Seven, Plus or Minus Two' in which he observed that humans can generally only remember seven 'chunks' of information at once[vii]. However, as we get more familiar with a type of information, the size of each 'chunk' can increase. The word 'invisible' might be one chunk to you and me, but a toddler just learning to speak might see it as four chunks, 'in-vi-si-ble'. A beginner chess player will look at each square individually, whereas a grandmaster's eyes will dart between groups of squares. A mathematician, on seeing the sum '12 + 15', may immediately process this as 27 without having to break it into 10 + 10 + 2 + 5. Memorization allows us to increase the amount of information carried in each 'chunk', a process Miller called 'recoding', and set us up for more complicated work.

Consider an eight-year-old learning her times tables, and repeating them parrot-fashion just as she might have done in Victorian England. *Five fours are twenty; six fours are twenty-four.* Her teacher hopes that this repetition will make the material 'stick' in her brain. She may not yet have learnt *why* five fours are twenty, but so long as the memorization is accompanied by other exercises that promote conceptual understanding, the exercise is not in vain. Knowing one's times tables by heart may make it easier for the underlying concept to reveal itself. In Willingham's words, the factual knowledge precedes the skill. Eventually, the idea will be illuminated: perhaps one day she will think of 'five fours' as an arrangement of twenty dots in a five-by-four rectangle. When this happens, we can be confident that she has learned the concept.

The example of the girl leads us to think of learning as split into two key components: one technical, and one adaptive. The technical part is memorization. Most of us can be confident that enough repetition of an idea will keep it in our memory, at least in the short term. Progress is visible: we can instantly check whether the girl has memorized her times tables correctly. As we have argued, this piece of the learning puzzle should not be overlooked: memorization acts as the scaffolding upon which more meaningful learning can be built.

However, Dan's eventual aim is for his students to *understand* the material. This is the adaptive part of the challenge. Although memorization is an important step towards learning, the real benefits come when concepts are understood. Scaffolding is useful only for the purpose of helping to build the underlying structure. Einstein is supposed to have remarked that "education is what remains when one has forgotten everything they learned at school". He was hinting that understanding is the more profound part of learning that is built on top of memory. Once the learning has happened, the scaffolding of memory can come down.

The transition between memorization and understanding can happen suddenly or gradually. The goal of the teacher is to facilitate this transition, even if none of us can see it happen. Understanding is mostly invisible: we can try to measure it by asking searching questions, but it is a subjective idea. It is hard for us to know *ourselves* how well we understand a concept, let alone for it to be measured by someone else. Exams and tests do their best to make progress visible, but most do a poor job at distinguishing memorization from understanding. In the real world, there are no dollar signs or lightbulbs appearing above our heads when we master an idea. Most learning is, unfortunately, invisible, and this is part of what makes teaching so hard.

TEACHING ≠ LEARNING

Heifetz argues that to 'exercise leadership' is to get people to confront an adaptive challenge[viii]. This is a shift from how most think about leadership in the professional world. Most teams have tasks to complete and a 'leader' to oversee their progress. If these tasks are technical challenges, then all they really need is *management*, not leadership. To compile the weekly report, the authority figure just needs to direct and organise the group. You can be an excellent manager, with top organisational skills and great respect from the team, without ever needing to exercise leadership. Most people only need to exercise leadership a small proportion of the time; many never need to do so at all.

If learning is an adaptive challenge, then to teach effectively is to exercise leadership. A teacher needs to get students to confront this challenge and raise their level of understanding. This is where Heifetz' framework is useful: as we shall see in Part IV, there are helpful recommendations for how to lead from a position of authority. A key insight is that authority is both a resource and a constraint[ix]: being at the front of the class, or being at the head of the team, comes with invisible expectations from all sides which may be difficult to fulfil. Counter-intuitively, it will often be important to *disappoint* those expectations if the class is to maximise its learning. For example, students might sometimes expect the teacher to 'give them the answer', but to oblige may prevent understanding from happening.

On the whiteboard in Dan's office, he has written in bold black marker the phrase 'TEACHING ≠ LEARNING'. He believes that most lecturers pay too much attention to what they are doing in a class. Planning is important, and professors are right to spend time in advance thinking about how the class will be structured. One downside, though, is that professors will often evaluate themselves against what they said they would do in the plan. If they covered the material without tripping up, and gave the right explanations at the right time, they will be happy about how the class went.

The problem is that this has no relation to what the students actually learned. Maximising learning across a classroom is a complex task, and requires a degree of flexibility on the part of the lecturer. Opportunities for student-to-student interactions may present themselves during class, and these may result in greater learning than if the professor had covered the same material herself. Some students might be on their phones during class and be oblivious to the professor's explanation of the concepts, however insightful they happened to be.

The failure here is an example of a common flaw in our approach to adaptive challenges: we try to solve them with technical solutions. Terrified of the unknown, we seek to replace problems we have no idea how to address with those we know how to solve. This is a form of 'work avoidance', a term coined by Heifetz[x]. If you have ever procrastinated by spending your whole afternoon making a brightly-coloured work plan, and then congratulated yourself on being productive, then you will be familiar with the concept. By focusing on what they know they can measure – the extent to which a lesson went as planned – some professors may avoid the harder work of trying to gauge the learning of their students.

Many of us have been in a lecture hall in which a professor asks, "Is everyone okay with this?", then treats the subsequent silence as an affirmation, even though deep down both professor and students know they are in trouble. To Dan, being a good *listener* is even more important as a teacher than being a good communicator. He measures the success of a class on the amount of learning they observe taking place. Since learning is an adaptive challenge, measuring progress is difficult, and it takes some experience to gauge the feeling in the room. It also requires the ability to 'get onto the balcony', to use another Heifetz phrase[xi], and observe the dynamics in the room with an impartial eye: this is hard to do while explaining concepts at the same time.

Fortunately, there is help available: tools such as Poll Everywhere[*] are an excellent resource to help teachers monitor students' level of

---

[*] Other polling tools are also available: Dan highlights Kahoot, which 'gamifies' learning, and Mentimeter, which allows users to create interactive presentations along with polls, quizzes and word clouds on the fly.

understanding during class. We will go into detail in Chapter 14 on the data that Dan collects to make the invisible visible. Taking the temperature using polling technology during class is not just a gimmick: it is an important evolution that deserves to become standard pedagogic practice everywhere.

RECOGNISING COMPETING PURPOSES

Although few would have much trouble identifying 'maximising learning' as the purpose of the class, there may be several competing purposes in play, all of which are invisible. For students, *maximising enjoyment, minimising effort, maximising test scores, maximising learning about something else,* and *making friends* are all likely to be found across the classroom. For the teacher, we can add to that list *maximising evaluation scores,* which may be closer linked to student satisfaction than to student learning.

These competing purposes should be recognised, and not ignored. The goal is always to maximise learning, but the presence of the competing purposes affects the strategy we should adopt to achieve that goal. Teachers have two choices in dealing with competing purposes: they can try to *use* them to their advantage, or they can try to *push through* them. They are like the captain of a sailboat being pushed away from its destination by the wind. The best solution is to use the wind to the boat's advantage by adjusting the sails. Sometimes, this is not possible, and the only way to make progress is to turn the motor on.

Competing purposes arise in the classroom because students are human. Robots, when they are programmed, can do their learning directly: they will move quickly towards their goal like an object pushed in space with no other forces acting on it. Trying to make learning 'fun' for a computer will just slow it down. In contrast, humans will be pushed around by the multiple purposes around them. If we try to take the same approach on the boat, pointing it towards the island and hoping for the best, we may be pushed away by the wind, or make slower progress than is optimal.

Maximising enjoyment

All students want to enjoy their classes. This simple truth means that teachers who can make learning fun will, other things equal, be more successful. If the only way students can enjoy themselves is to ignore the class and play on their phones, then student enjoyment is a headwind that slows the boat's progress. By making learning fun, teachers can adjust the sails to take advantage of the wind and speed the boat up. The progress it makes may not be directly towards our destination, but we can use the sails nonetheless to pick up speed and get closer to it than we would otherwise have done.

The extent to which enjoyment is a priority will vary significantly across students. Some will see enjoyment as a secondary aim to learning; others will only be able to learn if they feel like they are having fun. This creates a dilemma for the teacher: adding fun to an exercise may distract the first group and focus the second. In part, this can be mitigated by strengthening ties between the enjoyment and the learning. To an extent, though, the dilemma is unavoidable. As we will discuss in the next chapter, "maximising learning" does not mean maximising the learning of each individual student. We are aiming to maximise the learning of the group as a whole: the students all have different forces acting upon them, but it is the sum of all these forces that determines the overall force acting on the boat.

The link between learning and enjoyment is something that has driven Dan's entire career. He tells a story about one of his graduate school professors in a theory-heavy math class. The professor was a brilliant mathematician, but the class was totally at sea in the math notation. Dan, like so many students, found himself sitting in class so lost that he felt he did not even know what question he could ask to help himself.

"I remember he said something like 'this should be completely obvious to you', and to me that was crushing... we had a conversation after the mid-term exam, in which the class had averaged 20 out of 90. It was 25 years ago, and I still remember what he said:

'Frustration is necessary for learning. This idea that you can enjoy learning is a very American idea'."

Dan pauses, the memory of that conversation etched on his face. "I felt so offended by that claim. My professor felt that to learn, you had to push yourself. One of my dreams was to go back to Venezuela* and start a university, and I vowed that I would write in the walls of the university that frustration was *not* necessary for learning."

Dan does not see it as his role to entertain: he believes that his purpose is to maximise learning, and everything he does is in aid of that purpose. But as we have seen already, he uses humour several times

---

* Dan was raised in Venezuela, and moved to the US when he attended university.

in every class. Humour strengthens the invisible bonds between him and the students, as we will discuss more in Chapter 6. It also serves to release tension building up in the class caused by the cognitive work that learning creates, which we will look at in Chapter 13. When I ask Dan *why* he uses humour so much in class, I am expecting him to point to one of these dynamics: usually, everything he does in class is calculated. His answer surprises me:

"Honestly, I use humour because I feel joy when I'm teaching. When you feel joy and you relax, you crack jokes."

The use of humour in class is spontaneous, and comes from a genuine love for teaching. But Dan is aware of the positive side effects that this can have on the class. He continues: "The most important aspect of using humour is to be authentic. Don't try things that don't come naturally. I'm not a funny person, but I try to use the energy of the classroom to make something more memorable, and connect at a human level. I say things like 'that's a nice tie you're wearing today'. People laugh because it's disarming: professors aren't always making those kinds of comments. It connects at a more human level with the class."

The point about being authentic is important because it implies that there is no 'correct' way to use humour in teaching. Authenticity builds trust: trust and humour together are a powerful force to strengthen the invisible bonds between Dan and the students. But we should also look at Dan's answers with a sceptical eye. It may be that in indulging his joy of teaching and connecting with the class, he occasionally veers off towards this competing purpose of maximising enjoyment over learning. His visceral reaction to the phrase *"frustration is necessary for learning"* is fascinating: his own students will go through hours of frustration in his problem sets, even though they are designed to be relevant and engaging. My own suspicion is that when he reiterates to the students that his central purpose is to maximise their learning, and not their satisfaction, he is also doing that to focus himself.

We said before that by making learning fun, teachers can use the sails to take advantage of the winds acting on the boat, and make more progress towards their goal. The second way to address the issue is to *focus* the students, turning on the motor to push the boat on through. A focused class will be less insistent on being entertained, but just as the boat has a limited supply of fuel, the teacher must be selective in when to apply focus to the class. We have already seen one example of this in the Mexico case in Class 4, when Dan told the class that he wanted a high level of energy in the room.

In the discussion about Dan's opening class, we spoke about the idea of a contract: a vision for the course that both teacher and students sign up to. As well as strengthening the invisible bonds between Dan and the students, this contract is designed to help focus the students on learning, rather than any other competing purposes. As we will see later, he also makes it explicit that his purpose is to maximise the students' learning, not their satisfaction. Acknowledging that students have competing purposes is the first step to managing them.

Minimising effort

This is a tougher competing purpose to deal with. While a good teacher may design a course which makes learning fun, no-one has yet found a way of getting students to learn with minimal effort. Minimising effort is another purpose shared by all students: other things equal, we would prefer to make the same gains with lower effort, and use the spare capacity for something else.

The strength of the headwind depends on their levels of motivation. Highly motivated students will tend to prioritise maximising learning over minimising effort, and vice versa for those lacking motivation. Motivation is linked to enjoyment: to varying degrees, students are more motivated to put in effort when they are having fun. But it also depends on many other factors: their level of natural interest, perceived future benefits of the course, workload from other courses, and personal circumstances. Here is another example of the invisibility of learning being a hindrance: not being able to quantify the future benefits of learning means motivation is weaker than in the fantasy world we outlined at the start of the chapter.

Depending on the reason for low motivation, there are different ways of focusing the class. Firstly, for students without a natural interest in the subject, a teacher's enthusiasm can help to inspire, pushing the boat through the headwind. Subconsciously, students may think "if a person I respect loves the subject, then maybe there's a good reason for that."

This is clear in Dan's approach to teaching. At the start of the first class, he tells them that he is standing in front of them because he loves to teach. His passion for the subject is written on his face and demonstrated by his high energy levels. This is not to say that a teacher needs to be energetic to be passionate: Dan points to Brian Mandell, who teaches advanced negotiation skills at the Kennedy School, as someone whose passion for the subject is never in doubt, but whose sense of humour is deadpan and sarcastic. The two have different teaching styles, but both are authentic. Students trust that the professors' *purpose* in life is to love and to teach their subject: they are not just doing it for the money. Passion and purpose are tightly linked.

Secondly, for students who are not convinced the class will be of great value to them in the future, it is critical to motivate each topic covered with relevant real-world examples. We saw this in the case of Bayes' Theorem, where the discussion around the advantages and disadvantages of mammograms motivated the technical exercise that followed, and then with the Mexico case study. The sizeable amount of work required to understand a new statistical concept will only be done by students if they think it is worth it to them.

Thirdly, students who are struggling with the workload from other courses may be inclined to lower their effort to compensate. To some extent this is inevitable, but can also be mitigated by working with other teachers and professors to manage the combined workload over term. Dan works closely with the head of the program, and the other professors teaching it, to ensure that weeks that are particularly 'heavy' do not coincide with similarly demanding weeks in the other courses.

Finally, students may come into the class affected by personal issues, or other problems not related to the course. This cannot be avoided, but can be mitigated. As we will see in Chapter 13, lowering the 'heat' in the room may help a student under stress to focus on learning. In 2020, teachers had the challenge of maximising learning while many of their students were worried about the effects of the coronavirus pandemic. Many adapted their style accordingly, and aimed slightly lower in the amount they expected students to learn: maximising learning subject to severe constraints requires acknowledgement of those constraints.

In some teaching roles, especially in primary schools in which teachers are responsible for the same children every day, a teacher may also take advantage of the strong bonds between them and the children to help them through personal issues or other challenges. Here, the dilemma between individual students and the wider class is at its most testing: teachers may feel emotionally compelled to spend a large proportion of time supporting a child in need, at the expense of the learning of the rest of the class. In many real-world situations away from Harvard, supporting kids through childhood becomes a goal more important than maximising learning.

Maximising test scores

A well-known barrier to understanding is the incentive to optimise exam performance rather than underlying learning. In a previous life as a math tutor, I had a student who memorised the quadratic equation formula: ask him to solve for x in $3x^2 - 7x + 12 = 0$ and he would have no trouble at all. However, if you asked him to solve $x + 1 = 0$, he was unable to conclude that $x = -1$. The student knew how to plug in the numbers to the formula, but did not have any of the understanding of how the equations worked. In the real world, this would be useless. For his exams, it worked.

What we see here is what Heifetz would call a *technical solution to an adaptive challenge*[xii]. The student had focused on memorization ahead of understanding because this would get him through the exam. But learning is an *adaptive* challenge that requires deep changes to occur in the brain. Memorization will not, on its own, bring about those changes. Technical solutions to adaptive challenges do not tend to last for long.

Like the other competing purposes, there are two main ways that this dilemma can be addressed. We can either use it to our advantage, or push on through it. To achieve the first, we need maximising test scores to get us closer to the goal of maximising learning. This is done in the design of the test, ensuring that it rewards understanding and not memorization. This is difficult, because questions that reward understanding usually require written prose that is harder to mark, and more subjective.

Dan and his course assistants spend an entire day marking students' exams, debating among themselves the extent to which answers have shown understanding, and trying to ensure consistency. This works in his class of 80 students, in which he sets the exam questions. It is much tougher to do at the high school level, where exams are standardized across entire countries and marked by thousands of different people. This raises the importance of objective marking, and gives teachers a serious dilemma when deciding whether to maximise learning or maximise exam performance.

Alternatively, we can try to push through it by getting the students to focus on learning rather than test scores. This, too, is much easier in a graduate university setting than at high school or undergraduate level. For most of Dan's students, the exact grade they achieve in the exam will not be relevant to a future employer. Only a few have failed the course, and they are given the opportunity to retake it the following year. This is all in contrast to high school exams, in which the exact grades will often be used to inform university entrance offers and early career prospects.

Although the reasons above mean that this competing purpose is stronger in schools than in universities, there is one important principle common to both. The temptation to prioritise memorization over understanding is strongest the night before the exam, and weakest at the start of the course. Being aware of this, teachers can start the course focused entirely on promoting understanding, then build in some elements of memorization as the exam gets closer. As the captain of the boat, if you know a headwind is going to pick up later on in the day, you aim to make as much progress as you can while the wind is behind you.

With these issues in mind, Dan has a unique way of tying exams into the learning process. After attending a workshop run by the Nobel Prize winner Carl Wieman, he was converted to the idea of *two-stage exams*. The first stage of the exam is the same as any other. The second stage brings the students together in groups of four, and gives them all a subset of the questions they just answered in the first stage. The students are encouraged to discuss their answers in their groups, and they all submit a second version. Groups who manage to improve upon the students' individual scores get a small boost in the final outcome.

Dan reflects on what piqued his interest in this new type of exam. "I was often frustrated that the students did not reflect on the exam: it seems like such a wasted opportunity for learning. We found in a normal midterm exam that only 38% of students downloaded the solutions that we made available afterwards. In my office there's a crate in front of me with exams from another professor, and those exams stay there forever. The students don't pick them up.

"It's an ideal opportunity for learning: they prepared for the exam, they tried to apply their knowledge, and then they get no feedback on what they learned, other than a grade. We *know* that feedback is important for learning, and this stage of the exam process is less than ideal. Students see the grade, and that's it. Two-stage exams give the notion that the exam is not just about the grade, it's deeper than that."

Dan's use of two-stage exams kills two birds with one stone. Firstly, he maximises learning by ensuring that the exam itself is a learning experience. Second, in doing so, he makes clear that the grade is less important than the learning. Two-stage exams have not yet 'taken off' around the world, and grades remain the key outcome of most exams for most students. Dan, though, has taken advantage of his position in a graduate university environment to push the idea forward.

Exams also give us a concrete example of Dan's attempts to promote understanding over memorization. He encourages students to bring with them a 'cheat sheet', which can include anything they want. He does not want students being tested on whether they can remember the formula for the standard error of a distribution, and does not believe that it will be helpful for them to do so. Allowing cheat sheets means that students can bring more than just formulas with them: if they were struggling with a topic, they could bring all sorts of pre-cooked explanations to help them with the exam questions. In turn, this acts as an extra incentive to Dan to set questions that rely on a deeper understanding.

The last, and most important, way to move students away from test scores is to give them a greater purpose. In the next section, we will talk about the importance of inclusivity in the learning environment, and argue that learning is maximised when the class acts as a team to support each other. One student sums this up nicely in their evaluation comments:

"The inclusive atmosphere of the class was the key driver of allowing the class to be eager to learn. The focus really shifted away from grades and points, and towards understanding concepts and learning."

The key link is between inclusivity and the shift of focus away from test scores. When students are given the responsibility to support each other, their own score falls down the list of priorities.

Maximising learning about something else

We discussed earlier the varying tendency of students to want to minimise effort. One of the reasons this headwind might be strong is if a student does not have a natural interest in the subject. They may be taking the class because they have to, or because they think it will help them to understand something they *are* interested in. The goal of these students will not be to maximise their learning about the subject being taught, but to maximise their learning about the thing they are interested in. Performance will vary throughout the course depending on whether the student perceives the topic to be relevant to their interest.

The teacher's response will depend on how widespread across the class this other interest is. If half the class is interested in soccer, it may be a clever idea for a math teacher to use soccer-themed examples to illustrate key concepts. However, it is probably a bad idea to focus *only* on soccer given the alienating effect for the rest of the class. A mix of topics to reflect the diverse interests of the class will work best.

In Dan's case, the interest in policymaking is so central to the students' presence at Harvard that it would be foolish to run a statistics course that did not acknowledge it. He goes as far as to build this into the purpose of the course: as we heard earlier, his purpose is "not just to maximise learning about statistics, but also to maximise learning of the skills that will be useful to have out there in the world." The destination of the boat is a place where *useful* statistical learning has been maximised. By defining his purpose in this way, he ensures that the students, despite the initial fear of a class named 'Advanced Quantitative Methods', are aligned in their purpose for being there.

My suspicion is that the success of *this* statistics course over others is linked to the fact that this dual purpose of practical policymaking is built into it. Dan is never short of examples to motivate the students and help to illuminate concepts. This is partly his flair for teaching, but also good luck at being situated between statistics and a practical field. Courses that treat statistics as a branch of mathematics do not have this natural advantage, and often leave their students wondering why on earth they need to learn the formulas in front of them.

Making friends

Friendships are an integral part of the school and university experience, and it would be a sad world in which the joy of meeting other people did *not* distract us from work. Here again there is a distinction to be drawn between the different levels of education: the younger the child, the more importance is attached to personal and social development. By the time students come to Harvard, this development may still be happening, but is not usually seen as the responsibility of the professors*. Making friends becomes more directed towards professional rather than personal development, though the joy of doing so endures through life.

Once again, the goal for the teacher could be to 'use the sails' to bring the competing purpose of making friends closer to the main one of maximising learning. As we will see in Chapter 6, in many cases the two purposes may be perfectly aligned: one of Dan's key strategies to maximise learning is to foster student-to-student interactions in service of the learning. When it is possible to do so, Dan believes they will learn more from each other than from him: we saw this in the early classes with the way he put Juliana and Carla's pictures and quotes up on the board, and asked them to explain them. Class engagement improves significantly: the excitement of seeing their friends' faces on screen, and the anticipation of what they are going to add to the discussion, create a perfect environment for learning for a few important seconds.

There will be times, though, in which making friends does not support the learning of the class. The students at Harvard tend to be committed and motivated, so examples of this tend to come at the beginning of class while conversations have not yet died down, or students arrive late because they are still finishing their lunches together. This is where the contract from the first class, combined with enforcement of norms, should have some impact. The students are

---

* Ron Heifetz' class on Adaptive Leadership is an exception: many come out feeling they have changed as people, are more able to understand the hidden emotional dynamics of those around them, and feel more comfortable building bridges on controversial topics like race which may otherwise act as barriers to friendships.

asked to be "in their seats and ready to learn" by one-fifteen, putting the emphasis on the work of learning. It does not always work.

Maximising evaluation scores

In our fantasy world, professors would be evaluated based on the amount that students have learned. In the real world, this is not always the case, as illustrated by a fascinating study in 2019. Louis Deslauriers, a Harvard physicist, found that students who were asked to participate actively in the classroom *perceived* themselves to have learned less than students to whom the professor gave a more traditional lecture with the students in a passive role[xiii]. They also enjoyed the traditional lecture more, gave better evaluations to the instructor, and were more likely to want all their courses taught this way. Based on subsequent test scores, they had actually learned *more*.

We should caution that these tests were restricted to college physics students: it is possible that there are traits among this subgroup that lead them to be more comfortable learning passively than actively. In either case, the paper shows that it is possible for students to misperceive their own learning. In particular, Deslauriers found that some students "dislike being forced to interact with each other", "resent the increase in responsibility for their own learning", and complain that "the blind can't lead the blind".

This is a good example of the principle, highlighted earlier in the chapter, that it is sometimes necessary to disappoint student expectations to maximise learning. There is a simple reason for the misperception of the students: *they associate cognitive load with failure*. Let's imagine you were a top art student in class: your whole life, you have been told your paintings are wonderful, and the talent comes naturally. On the other hand, you hated math: none of it ever seemed to 'click'. Your brain struggled to cope with the cognitive load during math class, but the hours you spent painting just seemed to fly by without a care in the world. You end up associating struggle with failure, and relaxation with success.

Figure 6.1: active learning increases performance, but decreases satisfaction

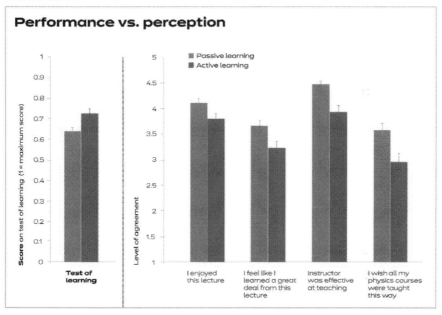

It is the same for the physicists, who have spent most of their life being very good at physics. The higher cognitive load involved in active learning is a good thing: the more work the brain is doing, the more physics they are learning. But it feels like they are not getting it, and they blame the instructor as a result. This has important implications, as Deslauriers highlights[xiv]:

"Attempts to evaluate instruction based on students' perceptions of learning could inadvertently promote inferior (passive) pedagogical methods... a superstar lecturer could create such a positive feeling of learning that students would choose those lectures over active learning."

The competing purpose of maximising evaluation scores may therefore lead a teacher to give less responsibility to the students than they ideally should. Without realising it, their lecture style is giving them the answers. Students like this because it means they feel like they are 'getting it'. Often, they are not.

Another area where evaluations and learning can be in conflict is in the length of problem sets. Every year, students complain about the length of the homework they need to do for Dan's class. Dan holds firm, and tells them: "If my purpose was to maximise evaluation scores, I'd make the problem sets shorter, but my purpose is to maximise learning."

The fact that Dan's evaluations are so strong – up above 4.8 out of 5 for the last five years – is a counter to this idea: if he is really sacrificing evaluation scores for learning, why are those scores so high? One answer is that by making clear to the students that his purpose is to maximise learning and not evaluations, he focuses the students towards that purpose. He motivates them to remember they are there to learn. When it comes to the end-of-term evaluations, their purpose has moved closer to his. Additionally, the fact that the problem sets do so much preparatory work for upcoming lectures mean that the students reap the benefits of their struggle in the lectures themselves.

The students can feel the success of active learning, and they stop associating cognitive load with failure.

With so many competing purposes in play, it is difficult for the teacher to keep the boat steady, while moving quickly towards its goal. It is a delicate mix of adjusting the sails where there are opportunities to use the winds in their favour, and turning on the motor when needed to push the boat on through. In Chapter 13, we will look at the risks of using too much fuel at once, and overheating the engine: the learning environment may be damaged if the professor forces the students to do more cognitive work than they can collectively bear. Before we get to that, though, we should tie down the nebulous idea of the 'learning environment'. Defining it in more concrete terms will be key to understanding the invisible dynamics of the classroom.

6

THE FOUR TYPES OF LEARNING ENVIRONMENT

Although thousands of students have found Dan's class transformative, few can articulate why they learned so much. One reason it is hard to explain his success is that the concept of the 'learning environment' is an invisible one. We can speculate about how his humour contributes to a positive learning environment, but without defining the concept, we have no way of knowing whether this is true or meaningful. It is hard to put into words the warmth that one feels in the presence of a teacher that supports us through our learning. The goal of this chapter is to try.

Ron Heifetz, writing in *Leadership Without Easy Answers,* develops the idea of the 'holding environment'[xv]. Heifetz's background was in psychiatry, and the term originates from this field in describing the relationship between therapist and patient. While our patient may be going through the adaptive challenge of a difficult break-up, his therapist 'holds' him through the challenge by protecting him while trying to make progress on the challenge. Heifetz extends the idea of a holding environment, defining it as "any relationship in which one party has the power to hold the attention of another party and facilitate adaptive work".

In this book, we define the 'learning environment' as the holding environment of the classroom. This includes all relationships between all the members of the classroom: the teacher, the course assistants, and the students. The learning environment is a network: one can imagine invisible bonds joining each person in the class to each other. When these bonds are strong, the learning environment is strong. When they are weak or non-existent, the learning environment is usually weak. We distinguish between *vertical bonds*, which describe the relationship between teacher and students, and *horizontal bonds*, which are the relationships between the students themselves[*].

We saw a small example of a strong holding environment in the way Dan poked fun at John, the American student who speaks like he is being interviewed on CNN. If the vertical bond between Dan and John had been weak, a joke at his expense could have severed it: a relationship that starts badly is difficult to recover. If the other vertical bonds between Dan and the class had been weak, they might have felt defensive about the way their friend was being caricatured. If the horizontal bonds between the students and John had been weak, they would not have felt so comfortable laughing at his misfortune, and the

---

[*] In classrooms where there are course assistants, there are additional vertical bonds in both directions between them and the teacher, and between them and the students. There are also horizontal bonds between the course assistants. In classes where there is more than one teacher, the horizontal bonds between teachers are also highly relevant to the learning environment.

joke would have increased the tension in the room, rather than released it.

Figure 6.1: illustration of the learning environment, comprised of the invisible vertical and horizontal bonds between members of the classroom

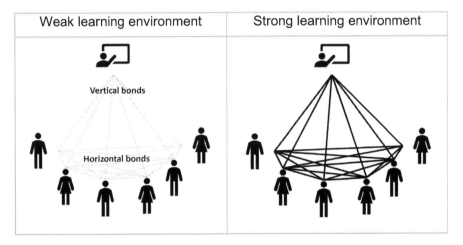

Instead, all these bonds not only *withstand* the joke; they are *strengthened* by it. The process is the same as when friends get acquainted: after an initial period of politeness, there is a moment that someone tries a sarcastic comment or a joke at the other's expense which cements the relationship. Trying the same comment as an ice-breaker the first time they meet might have the opposite effect.

Speaking to Dan about this moment in class, it is interesting to see his discomfort. He often brings it up in our interviews, worried that the reaction of the class could have been different, and he is keen to stress that jokes at the expense of students may also be an indicator of an unhealthy approach by the professor. "The more I think about it", he says, "the more I feel like I've been speeding on the highway without an accident. That thing about CNN was in one of the first classes of the semester. I need to think more about how to do it."

I think he underestimates the instincts that were in play in the moment: like a stand-up comedian, he could sense that the class was warmed-up enough for a remark that poked fun at one of them. He played off their laughter in a way his instincts would have warned against in a different scenario. His concern since that class is an indication, though, of how carefully he cultivates the learning environment. Even though the joke went down well, anything with even a risk of damaging the vertical bonds with his students is something that will keep him up at night. For better or for worse, he wants his students to like him.

He need not have worried. John, a year after the class, still remembers the moment for the right reasons. "My reaction at the time was definitely one of amusement", he says. "Light humour, and teasing, is a way to cultivate a more intimate and open setting. With Dan, every time you walk into his class, you fell as if you're immersing yourself in an experience, which demands you to be awake, to be alive, and to participate in the friendly and playful vibe of the class. Through his humour and self-deprecation, he's able to create this familiar setting where people really feel at ease."

The point about Dan's self-deprecation is an important one. Dan's humour tends to bring himself down to the level of the students: he may be the professor, but he is also one of them. Everyone calls him 'Dan' in class: to say 'Professor Levy' instead would feel formal and out-of-place. As John highlights, this makes it easy for them to accept and take part in the jesting that happens within the class. Professors who mock their students from an elevated position may be more likely to be interpreted as bullying, which will harm vertical bonds.

It is not just the responsibility of Dan and the students to build these bonds between each other. Schools and universities will do everything they can to ensure the horizontal bonds between students are as strong as possible before they enter the classroom. This is certainly the case in the program* these students take, which starts with a three-week 'Math Camp' before term starts. That may not sound like fun, but there is a deliberate emphasis on extracurricular activities, which are just as important as the math itself. By the time Dan's class starts, these students have all been on nights out together, and spent hours in each other's company. This allows Dan to build community in his class more quickly.

Given our central purpose of maximising learning across the class, we are interested in how professors like Dan can create an environment that will do so. Based on the definition of a learning environment of being composed of the horizontal and vertical bonds between the members of the classroom, we define four separate types of learning environments: *non-interactive, interactive, inclusive,* and *team learning.*

One contention of this book is that 'team learning' is the most effective of the four because of the way it takes advantage of all the bonds in the classroom as conduits for learning. The four environments differ by

---

* Most of the students Dan teaches are part of a wider program known as the MPA/ID: the Masters in Public Administration in International Development. The MPA/ID teaches students economics with a practical spin, aimed at giving them the tools they need to change the world for the better.

the extent to which a teacher aims to strengthen the vertical and horizontal bonds of trust, as shown by the diagram overleaf.

Figure 6.2: the four types of learning environment

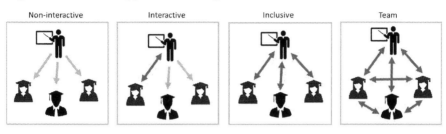

Non-interactive learning is the old-fashioned model: the teacher teaches, the students listen. Once teachers realise that they are expected to involve the students in some way, they start to ask questions of the students. This makes the learning *interactive.* But what often happens is that the same two or three students answer all the questions, which is not helpful for the learning of the whole class. This describes a great deal of university experiences across the world.

Teachers who are deliberate in their attempts to involve the whole class are seeking to graduate to *inclusive* learning. We will discuss some of the tools and techniques that Dan uses to achieve this. The goal of the teacher is to maximise learning across the class, while the goal of individual students is either to maximise their individual learning, or one of the competing purposes (enjoyment, test scores, making friends, etc.) mentioned in Chapter 5.

At the top of the pyramid is *team* learning, which comes about when the purpose of teacher and students is aligned: both are seeking to maximise learning *of the group*. The classroom acts like a team in search of its common goal, and the students take responsibility for their own learning. With teacher and students all pushing in the same direction, there is a greater chance of learning being maximised across the group. We go into more detail below on the four approaches.

NON-INTERACTIVE LEARNING: ARE TED TALKS EFFECTIVE?

The simplest form of teaching requires no interaction at all between teacher and students. A teacher can stand up, deliver the material, and leave. It might appear that non-interactive teaching is outdated and ineffective: we think immediately of the Victorian classroom from *Hard Times,* or the professor who reads out bullet point slides to students struggling to stay awake. This is not always true. TED talks, for instance, are non-interactive, but the best ones give us memorable insights. Learning can take place without interaction. Some might even claim that their success disproves the notion that teacher must interact with their classes to maximise learning.

If people are asked, 'what makes a great teacher', many of the answers will be compatible with non-interactive teaching. For example, we often hear that great teachers are 'charismatic': there is some overlap between traditional views of great teachers, and traditional views of great 'leaders'. However, as Deslauriers highlighted in his study, students sometimes perceive that they have learned more if they have a 'superstar' professor, when actually they have just been entertained[xvi].

A charismatic teacher may build trust with students quickly, strengthening the vertical bonds between them which make up part of the learning environment. But this will go to waste, or even be unproductive, if their goal is not to maximise learning. We trust charismatic figures because they entertain and reassure us. Even if we believe our sole purpose is to maximise learning, we all want to enjoy life and feel secure. So we trust a charismatic teacher, even if they are leading us down the wrong path. The boat may be travelling in the wrong direction, but its passengers, impressed by its progress, fail to notice.

There are obvious similarities between charismatic teachers and charismatic authority figures. This is no coincidence, given that a teacher is in a position of authority. Heifetz warns against the glorification of charismatic figures in *Leadership Without Easy Answers*, writing that "charismatic authority relationships can degenerate into mutual dependencies that erode critical judgment on both sides."[xvii] There is a thin line between trust and dependency: the former is an essential part of any relationship in which a service is provided, while the latter creates an incentive for us to protect a relationship even when it goes wrong. Mutual dependencies create strong bonds, but risk steering us away from our goals.

We can learn from TED talks, but not always in the way we think. The maximum length of a TED talk is 18 minutes: the founders, backed up by scientists, worked on the basis that the attention span of the average person does not stretch to any more than this. A great talk might give us a spark of inspiration, and a desire to investigate further, or to debate it with others. Dan believes this is where the real learning happens. "Can you learn something from a TED talk? Absolutely", he says. "But the learning is not happening during the talk, but afterwards. It happens when you go to your partner and tell them, 'can you believe what I just saw in this TED talk?', and engage in a conversation."

Dan points out that the science is behind him. He continues: "The learning is only happening because the brain is doing recall on what you heard. I don't deny that they can be so engaging that it provokes that conversation. That action, that reflection, is what forms the neural pathways to learning."

He is alluding to a book by James Lang, the American education writer, called *Small Teaching*, which goes in depth into the science behind how we memorize, and how we learn. Lang opens his book[xviii] with a discussion of the "retrieval effect", which says that "if you want to retrieve knowledge from your memory, you have to practice retrieving knowledge from your memory". On the neuroscience, he writes the following:

> *"Every time we extract a piece of information or an experience from our memory, we are strengthening neural pathways that lead from our long-term memory into our working memory, where we can use our memories to think and take actions. The more times we draw it from memory, the more deeply we carve out that pathway, and the more we make hat piece of information or experience available to us in the future."*

This is the best way to think about TED talks, and every other talk you've ever been to in which there was a good speaker but minimal audience interaction. They *can* inspire learning, and they are worth watching. But you should watch them with the knowledge that watching is not enough: you will soon forget that big insight. It is in engaging with the ideas that inspired you, speaking to others about them, writing about them, and asking about them, that the neural pathways are strengthened. A professor who presents an hour-long class with no interaction is failing on three counts: they are persuading students that they are learning when they are not, they are dramatically overestimating their attention span, and they are missing out on valuable class time to engage with the material and allow learning to take place. To maximise learning, plenty of interaction and engagement will need to take place.

INTERACTIVE LEARNING: THE RISK OF SUPERSTAR STUDENTS

Most university classes across the world fall into this second category. Professors are aware that attention levels will drop quickly if they do not have the chance to engage, and so they ask questions to check that the students are keeping up. There are two or three students in the whole class who have a true grasp of the material, and that's because they've already learned it once or twice before. Predictably, these are the students that answer all the professor's questions. Meanwhile, the other students in the class are secretly thankful that they have not been put on the spot because they did not know the answer.

The success of interactive learning of this type is that it can maximise the individual learning of those two or three students. Even if they are familiar with the material, being forced to explain it to the rest of the class is an excellent tool for cementing their learning. The most motivated and talented students get even better, and will be held up by everyone concerned as success stories for the class. They go off and have successful and fulfilling careers.

The big problem is for the other 90% of the class, who are learning little. They may end up with some learning at the end of term, but it will not happen in class: it will be through hard work on problem sets or through discussion with peers. They will always be playing catch-up, given the wasted class time. Their attention will drift through no fault of their own. They will start to browse on their phones instead. The 'interactive' element of the class is inaccessible to them: although they would be able to answer some of the questions if given enough time to think, they are always comfortably beaten to it by the class superstars. For them, the class is non-interactive.

This model is incompatible with the goal of maximising learning, for two reasons. The first is obvious: if only a few students at the top are learning then the sum of that learning, however you choose to quantify it, will be a sum of three numbers, rather than thirty. A useful metaphor might be the 'jam-jar challenge', in which you are tasked with scooping as much jam as possible out of 30 full open jam jars in the space of two minutes. Your best strategy is to take a big scoop out of each jar, rather than seeking to empty one or two jars totally. For those few students that do interact, you are getting diminishing returns by continuing to focus on them. You can only learn the same thing once.

This leads us to the second reason: if you want to maximise learning, then selecting the students at the top of the curve to focus on is your worst possible strategy. They have already learned much of what you are planning to teach: their jam-jars have been partly emptied to begin with. If any students are worth more focus than the rest of the class, it is those who are closer to the bottom. One student who struggled in class wrote the following in their evaluations, suggesting that their experience had previously been that those at the back were left behind:

"Dan is the only professor I have had in my lifetime who was genuinely interested in including as many people as possible into the learning experience, especially those who are feeling lagged and falling behind course material."

Education has sometimes been called the "great leveller", after the nineteenth-century American public school pioneer Horace Mann called it "the great equalizer of the conditions of men"[xix]. The idea is that by giving some form of free education to all, those who are worse-off will have access to the skills they need to prosper. It has not always turned out this way: the rich can afford better schools, and differences in resources and bandwidth can lead to widening social gaps. The same is true in the classroom: there will usually be reasons other than 'natural talent' which set the top students apart from the rest. The fact that so many classes fall into this model of being *interactive* but not *inclusive* exacerbates the problem: the best-prepared students prosper at the expense of everyone else. By taking a more inclusive approach, teachers can both maximise learning, *and* make the classroom system fairer to all.

INCLUSIVE LEARNING: SILENCE, WAIT TIME, AND PARTICIPATION

Since the standard way of creating interactivity – asking questions – is biased towards those who already know the answer, teachers should find ways distribute the work across *all* the students. The question-and-answer format, as commonly applied, denies 90% of students the potential to learn because they are quickly given the answer without the opportunity to do any cognitive work. As we saw with the jam jars, if we want to maximise learning across a class, we are maximising a sum across students. Given that each student only learns a concept once, and our goal is to maximise the number of concepts learned across students, we have diminishing returns for individual students.

Normalising silence

One easy change, which Dan is a master of, is using silences to his advantage. He knows which students are the strongest, and which do not participate often. Upon asking a question, he can wait until a sizeable number of hands have been raised. We saw earlier how he normalised silence in the first class: he wants his students to know that silences are okay.

Just as we associate cognitive load with failure based on past experiences, we are prone to making the same link with silence. In many classes, silence is an indication that both participation and understanding are lacking, and it can frustrate teachers. In Dan's class, silences are opportunities. Each silence acts as a subspace in which students can engage with a question and learning can take place. To curtail this learning by going to the first hand in the air is an opportunity wasted. This link between silence and inclusivity is not an obvious one: more naturally, we think of silence as an important rhetorical device that can add clarity to a teacher's explanations. Dan tries to do this too, but does not believe he has the gravitas of some other teachers. But when silence is placed between a question and its answer, it is a powerful source of learning for all.

The monk Thomas Merton beautifully articulated the different forms of silence[xx]:

"Silence has many dimensions. It can be a regression and an escape, a loss of self, or it can be presence, awareness, unification, self-discovery. Negative silence blurs and confuses our identity, and we lapse into daydreams or diffuse anxieties. Positive silence pulls us together and makes us realize who we are, who we might be, and the distance between the two."

Dan is seeking to create Merton's 'positive' silence, which focuses rather than blurs. At teaching seminars, he underscores the importance of silence as a tool: one professor introduced himself to Dan a few years after a seminar he took, and said: *"you are the guy who says silence is important"*. It struck Dan that of all the things he spoke about at the seminar, this was the one that stuck with the professor.

Mary Budd Rowe: the study of 'wait time'

The power of silence in ensuring inclusivity was not discovered by Dan. It has been understood since the early 1970s, when the American educator Mary Budd Rowe videotaped thousands of elementary school science classes over five years, and found that teachers allowed an average of just *one second* for students to answer questions, and would follow student responses with a comment in under a second. She writes[xxi]:

"When these two 'wait times' are extended to three to five seconds, a number of changes occur... there are increases in the length of the response, the number of unsolicited appropriate responses, student confidence, incidence of speculative responses, incidence of child-child data comparisons, incidence of evidence-inference statements, frequency of student questions, and incidence of

responses from 'relatively slow' students. The number of teacher questions which do not elicit a response decreases."

This is a remarkable finding from such a small change in teaching style. In the same study, Budd Rowe also finds that increasing wait times improves *teacher* characteristics, especially in the flexibility of teacher responses. Her findings on 'wait time' have been used as a guide for school teachers for decades, but Dan points out that they do not appear to have influenced practice at universities in the same way. There is great opportunity for improvement.

Teachly: participation software

Another way Dan ensures inclusive learning, and one that he hopes teachers everywhere will follow, is via the use of an online platform named *Teachly*. Dan is one of the inventors of the software, which is used in the classroom to monitor student participation. The class assistant responsible for it will see a layout of the class, with all the student names and faces attached. When a student speaks in class, the assistant clicks on them. In this way, Dan can record who has spoken, and how many times.

Teachly also asks students to fill in student profiles: these are the ones that Dan reads obsessively before the first class. The details of the students are automatically cross-referenced against their participation, so Dan can see what type of students he is calling on most. For example, some teachers may unknowingly have gender or racial biases in who they call on. Some may rarely call on students without English as a first language: this could be an indication that their 'wait time' is not long enough after asking questions. The students for whom Spanish or Mandarin is their first language are just as capable as the others, but may take a couple of seconds longer to process. One common bias is for teachers to focus on the centre, or the front, of the classroom, and ignore the sides and the back. Teachly gives the teacher all this information so that if biases start to creep in, they can be quickly addressed in future classes.

Dan, a data nerd, finds this all quite exciting. I remember the first time I came to his office for a team meeting with him and the course assistants. There was some anticipation to look at "how he did" in the first lecture in avoiding biases. He was calling more on women than on men, but the sample size was small enough for this not to be a worry. What he *did* need to address was that he was calling less on students without English as a first language, showing up on Teachly as a big red bar stretching to the degree to which the numbers are less than expected. He changed this in the next class: getting instant data meant that he was able to correct biases on the fly.

Far from being worried about being monitored, the students seem to take Teachly in their stride. They like that Dan cares enough about them to monitor these things. The monitoring is not tied to a participation grade, so there is no implicit threat that if they do not speak up, they will receive a weaker grade. It is more a tool for him: as well as bias correction, Teachly will flag the students who are speaking the most and the least. In class, Dan can try to bring in those who have not yet spoken by picking them for an 'easy' question, or finding relevant experience in the student profiles and contacting them in advance of class to get them to talk about it, or picking their response in the pre-class module to examine in class.

Several students in the post-course evaluations praise the high level of intentionality in everything Dan does, and Teachly is a prime example. There are no secrets here: the students know what he is doing, because he tells them regularly. Four examples of student comments are below:

> *"Dan is really intentional on bringing a diversity of participation from class, and encourages people with different professional backgrounds and general experience to contribute to form a well-rounded discussion."*

> *"The intentionality behind the structure, content, delivery of lectures and problem sets is astounding. A lot of work has been put into preparation of the course, and it has made my learning very enjoyable."*

> *"You can tell classes are tailored and very intentionally designed each time."*

> *"Nothing is left to chance – his instruction, the problem sets, and the manner in which he manages the classroom is all done with intention and care."*

The key takeaway from these comments is that students appreciate professors who put the effort into managing their participation. Intentionality is decoupled from the outcome: as we will see in Chapter 8 and elsewhere in the book, Dan sometimes gets it wrong. But just the fact that he is thinking about how best to manage *them*, rather than just managing the class material, makes them feel seen, heard and understood.

Warm calling

In bringing in students who struggle to participate in class, Dan uses a technique called 'warm-calling'. The name relates to the idea of 'cold calling', which is common at the Harvard Business School across the river, as well as the Harvard Law School across the yard. Teachers 'cold call' by picking out an individual student to answer a question. This strikes fear into the heart of students, and can be an effective technique, as that fear drives them to prepare for class in case they are called.

A 'warm call' still picks out a student, but gives them some warning. Dan will often say "I wonder if anyone would like to volunteer to answer this question", and move towards the student. If the student – let's call him Joe – has not put up their hand yet, he will follow up with "Joe, I wonder if you'd like to volunteer".

The difference between cold and warm calling is subtle but significant. Just as Mary Budd Rowe found that student answers improve in quality when you give them more time to think, Dan is padding out the response time for the student in question by pausing and asking for volunteers. By the time it gets to the student, they have had at least ten seconds to formulate an answer. As a result, he can still use calling as a tool to promote inclusiveness, while avoiding the fear that cold calling can create. One student writes that he "really appreciates how Dan pushed me to participate". A warm call is still a push, but students are more likely to feel they are being held through the participation process, rather than forced into it.

Another student writes the following:

> "Dan creates a welcoming and inclusive environment. I came to Harvard not used to speaking in class, and cold calling would make me panic. After the semester, even though I haven't participated as much as I should, I feel very comfortable in doing so."

The student's experience hints at a change in mentality that the student can take to other classes and life experiences. Being in a safe and inclusive space can build confidence that permeates other areas of a student's life. But we should still look with a critical eye: perhaps the student's admission that their participation was low suggests that cold calling could have been an effective technique to maximise their learning, if not their comfort. We will go into this more in the next section.

TEAM LEARNING: GIVING RESPONSIBILITY BACK TO THE STUDENTS

Inclusive learning is achieved when the teacher's purpose – to maximise learning – is pursued in a way that includes the whole class. *Team* learning is achieved when this is not just the purpose of the teacher, but of the students as well. They take responsibility for the learning of others. The classroom therefore works as a team, with a common objective, rather than individuals with separate objectives. The horizontal bonds of the container (those between the students) are strong in this learning environment. Students trust that they will teach, and learn from, one another.

The challenge is how to engender that sense of responsibility in students. There is a 'tragedy of the commons' problem to confront: a team approach might maximise the learning of the group, but each individual student's goal might be to maximise their own learning, possibly at the expense of others. We saw this with the superstar students who answer all the questions. In non-inclusive environments, this is the norm: competition wins over cooperation.

To confront this challenge, it is helpful to remember the following maxim, frequently repeated in Heifetz' leadership classes:

People only take responsibility for what they helped to create.

To get students to take responsibility for the learning of others, they must have been involved in creating the learning environment themselves. We spoke earlier about how putting up Juliana and Carla's quotes on the board made them feel like celebrities, and increased classroom engagement because everyone was interested in what their peers had to say. But the more profound effect comes from the combination of the quote and the photo: while they are on screen, it is *they* who are creating the learning environment. The photos cement the idea that it is Juliana and Carla's ideas that are defining the direction of the class.

Experiments run on the students are effective for the same reason. We have seen numerous examples already of this: the birthday game, the mammogram question; the choice of lotteries and the question on the population of Turkey. These all had concrete links to the class material: the first taught us that we are bad at estimating probabilities; the second that counts are easier to understand, the third that we are risk-averse with high amounts of money, and the fourth that we can fall victim to anchoring. But the deeper effect on the class is that they taught *themselves* these concepts by taking part in the experiments. The data did not come from some faceless study of decades past. It came from *them*, and if they had chosen differently, they would have learned something else.

One student writes the following in their end-of-course evaluation:

> *"The course made me feel included in the group. Thank you for fostering participation and showing me what a good dedicated teacher should look like. I leave with a lot of curiosity about pedagogic methodologies."*

The last sentence is fascinating: the student came away curious not just about statistics, but pedagogy too[*]. This is a natural consequence of the students being encouraged to teach each other, and the way Dan brings them into his calculations when he addresses trade-offs he is struggling with. The team learning environment means students feel responsibility for pedagogy in class, and it is no surprise that many come away interested to learn more about teaching techniques.

If the learning environment is strongest when the classroom is a team, then to maximise learning we will want to know the attributes that make a good team. Fortunately, a 2015 study from Google named *Project Aristotle*, which looked across more than 180 teams within the company and compared them across 250 attributes, has some of the answers.

---

[*] The author, of course, had a similar experience.

7

BUILDING PSYCHOLOGICAL SAFETY

The Greek philosopher Aristotle (384-322 BC) was the first to write the now famous maxim that "the whole is greater than the sum of its parts". This is the ultimate goal of any team: that by joining together in a common purpose, they can achieve a better outcome than if they were to go about their business individually. When Google commissioned the Project Aristotle study in 2015, they were alluding to this quote.

At the start of the project, many believed they knew what the results would be. Common sense dictates that the best teams will have a balance of skills and personalities. Julia Rozovsky, Google's lead researcher on the project, wrote[xxii]:

> *"We were pretty confident that we'd find the perfect mix of individual traits and skills necessary for a stellar team – take one Rhodes Scholar, two extroverts, one engineer... and a PhD. Voila. Dream team assembled, right?*

We were dead wrong. Who is on a team matters less than how the team members interact, structure their work, and view their contributions. So much for that magical algorithm."

In other words, the bonds between the members of the team are more important than the skills of the members themselves. Applied to the classroom, the strength of the learning environment will determine success, not the expertise of the teacher or the students. To the surprise of everyone at Google, individual traits did not seem to matter much. By far the most important predictor of team success was the idea of 'psychological safety': the belief that one will not be punished for making a mistake.

Psychological safety operates at the group level. This makes it a different concept from bonds of trust, which are between pairs of individuals. It is also different from simply 'being nice': in fact, too much 'niceness' can be a consequence of *low* psychological safety. In contrast, our willingness to be cheeky, or to joke at another's expense, only comes when we believe we can do so without negative consequences. Dan's poking fun at his students, as we saw in the first few classes, could be an indication of high psychological safety, and a strong learning environment.

That said, Dan cautions that poking fun at students should not be seen, on its own, as an indicator of high psychological safety. A professor who bullies his class will have achieved the opposite, and the jokes at the students' expense will weaken the psychological safety in the room. The comments need to be supported by strong vertical and horizontal bonds to ensure that students feel more safe, and not less safe, as a result.

Amy Edmondson: psychological safety pioneer

Harvard researcher Amy Edmondson was the first to introduce the concept of 'team psychological safety'. She first came across the idea in the early 1990s, when she was part of a team investigating medical errors in hospitals. The goal was to understand the factors that led to those errors, and subsequently to address them. When she first calculated the results of the study, she thought there must have been a mistake. The number of errors teams made was *positively* correlated with their effectiveness: the best teams made the most errors[xxiii].

This made no sense, until she had a "eureka" moment. "What if the better teams had a climate of openness that made it easier to report and discuss error?" she writes. "The good teams, I suddenly thought, don't *make* more mistakes; they *report* more."

Her 2016 book, *The Fearless Organization,* argues for the primacy of psychological safety in determining the attributes of successful teams in the workplace. Her work was cited heavily by Google in Project Aristotle.

In that book, she highlights that psychological safety has been shown to have a significant effect on learning and performance[xxiv], and cites research in neuroscience that shows that "fear consumes physiologic resources, diverting them from parts of the brain that manage working memory and process new information. This impairs analytic thinking, creative insight, and problem solving."

A big part of Dan's success is in creating a 'fearless classroom': one in which students are not afraid to speak up, and venture answers without complete confidence. We explore how he does this in this book, but for the science behind it, we have Amy Edmondson to thank.

We said above that the key to team learning is for teachers to give responsibility back to the students. Counter-intuitively, this can *increase* psychological safety. Rather than being fearful of the extra responsibility, the students are less afraid to question something that they created themselves through a process of trial and error. They have already experienced the ignorance that has been part of the struggle. Dan often encourages the class to explore wrong answers: ignorance is celebrated, not punished.

Edmondson describes in *The Fearless Organization* a concern that explains why employees often hold back on good ideas[xxv]:

> *"What if the current system is effectively the boss's baby? By suggesting a change, we might be calling the boss's baby ugly. Better to stay silent."*

In a class in which the material is all pre-determined by the teacher, students may fear questioning that material, just as we might fear criticising someone else's children. They may also not want to look ignorant in front of their classmates, and they may not want to highlight to the teacher that their explanation has not worked.

AVOID VALIDATING RESPONSES

An important technique used by Dan to promote psychological safety is to *avoid validating student responses* as soon as they are made. This means that 'wrong' answers result in no humiliation for the student concerned: Dan may later make clear what is right, but at a time when the class is not focused on any individual student. Instead of reacting to what the student said, he will ask the class if anyone has a different view.

He will also put responses back to the class when a student gives a 'right' answer. On one hand, he needs to do this to avoid the act of putting it back to the class invalidating the original student's response, but this is not the primary reason. When we talked about interactive teaching, we spoke about the perils of superstar students giving away the answers too quickly. Continuing the discussion after a correct response has been given prolongs the time in which cognitive work can happen in class. It also allows an opportunity for 'common mistakes' to surface and be discussed.

Dan confirms in interviews that this is an important part of his teaching. "There's a risk that you convert yourself into the validator of the students' answers", he says. "The goal is to help everyone learn. If you're just saying 'that's a great point', people aren't doing the thinking, they're just waiting for the instructor."

However, he is also wary of the trade-offs involved. "On the other hand, the risk is that anything goes. People say stuff that's unrelated. At the end of the segment, I do want people to know the right answer. You have to be careful to stimulate a debate, but without personalising, also signal what arguments have merit. I try to do it after a discussion has happened, and in a way that does not make anyone fearful of making a mistake in the class."

The path Dan seeks to navigate treads the line between the competing purposes of maximising learning and maximising satisfaction. When a good debate gets going, everyone wants to have their say. The teacher will often have to disappoint that expectation to maximise learning. As Dan points out, the key thing is to facilitate the discussion in a way that promotes psychological safety while being directed at the learning goal.

REDUCE FEAR OF PARTICIPATION

In the last section, we looked at Dan's preference to 'warm-call' students, rather than 'cold-call' them. This is also based on psychological safety. He explains: "My resistance to using cold calling is a consequence of not wanting to create tension and be confrontational. There's an aspect of that which serves me well, particularly with students that have had traumatic experiences with math."

That said, he admits that there is a trade-off. There may be times in which the safety the students feel in his class acts as a demotivator. "In creating that super-welcoming atmosphere... I think that if students had two assignments, for me and Brian Mandell*, and they could only do one, they'd do Mandell. I do see a tension between the two approaches."

He continues: "I think cold calling would make students come more prepared for class. On the other hand, I've had students who've told me that just seeing how much effort I put into teaching them makes them want to put in that effort in return. There's a human connection that they perceive: this guy's not trying to be mean to me. I want to

---

* The popular negotiations lecturer we saw in Chapter 5.

invite people into learning. I don't want them to learn because they fear humiliation."

The experience of one student typifies the appreciation that many of them feel, especially those for whom English is not their first language:

> *"I find it really valuable how Dan encourages class discussion all the time. As both an international student and an introvert, I felt really comfortable participating in this class even though sometimes I wasn't sure if I had the right answer."*

The combination of being introverted *and* not having English as a first language can be difficult for students coming to Harvard, where they are expected to contribute and to keep the pace with the native English speakers in the class. It will often be these students that Dan will be most focused on including.

Dan had a remarkable experience a couple of years ago that reinforced his belief that students should not be intimidated into participating. He tells the following story:

"I had a student who responded very well to one of the online modules. In class, I put his picture up on screen and called on him. He stuttered, and he explained, but it took a while. I went about my day without thinking anything of it.

The next semester, I had a faculty colleague who told me that this student had come to him and said that this moment had had a profound influence on his time at the Kennedy School. When asked why, he said that it was the first time in all his education – elementary, high school, college – that he had ever spoken publicly in a classroom.

It gave him the confidence to speak not only in my class, but other classes too."

Dan pauses, and reflects. "A cold call could have been a really different experience for him. At the end of the day, I'm not comfortable as a human being taking that confrontational approach."

This should not be surprising by now: Dan's nature is to avoid confrontation. While this generally serves him well, it may sometimes get in the way of his purpose of maximising learning. Perhaps a more confrontational style could put more responsibility on the students, and serve to *improve* psychological safety when they need to be told to do their homework. A soccer referee who firmly tells a player that one more foul will result in a yellow card will face less resistance when issuing that card later. Dan might feel safe telling jokes, but not in laying down the law.

In telling the story of the quiet student, he could be falling victim to the confirmation bias he teaches his students about: perhaps it confirmed his existing belief. It is possible that for every student encouraged to speak up by his warm approach, there were ten who failed to prepare for class. No approach should be taken as gospel: psychological safety and niceness are different things, and Dan may be missing out on opportunities by being too nice. Just as Dan encourages in his class, we can hold competing ideas in our heads.

Underpinning

The Google team at Project Aristotle concluded that there were *five* main traits of a successful team. Psychological safety was the most important, and their research suggested that it underpinned the other four[xxvi]. The five traits were as follows:

(1) *Psychological safety:* team members feel safe to take risks and be vulnerable in front of each other.

(2) *Dependability:* team members get things done on time, and meet Google's high bar for excellence.

(3) *Structure and clarity:* team members have clear roles, plans, and goals.

(4) *Meaning:* work is personally important to team members.

(5) *Impact:* team members think their work matters and creates change.

It is no coincidence that the five traits of successful teams are found in abundance in Dan's classroom. The bonds of trust between the classroom members develop when they take responsibility for each other's learning, and strengthen as they become reliable in doing so. This is the second key trait, *dependability*.

We have also seen the importance of the purpose of maximising learning being well-defined, given the other competing purposes in play such as maximising enjoyment. The learning environment is centred around this purpose. This is the third trait, *structure and clarity*. The students should know why they are there, and arrive ready to learn.

The fourth and fifth traits, *meaning* and *impact*, are difficult to generate out of nowhere, but we saw earlier how for those without an intrinsic motivation to learn statistics, they may still be motivated by the possibility of learning about something else. In Dan's case, the students are motivated to learn how to change the world through policy, and they need his class to do it. Meaning and impact come from outside, but they are also magnified by Dan. We have already seen how he deliberately motivates each example by telling the students why it matters to them.

The learning environment, which we defined as the collection of horizontal and vertical bonds between members of the classroom, is underpinned by psychological safety, strengthened by dependability, directed by structure and clarity, and motivated by meaning and impact. As we move back into Dan's class for Part III, we can do so with all these ideas in mind.

We now have a framework at our disposal with which to analyse Dan's teaching. We argued that learning is an 'adaptive challenge': progress is mostly invisible, so we can rarely be sure when or how it happens. Even afterwards, it is tough to pinpoint when the connections in the brain were made.

We highlighted the importance of having a clear purpose of maximising learning, and said that teaching and learning are different things. Using the sailboat as a metaphor, we also talked about the problem of competing purposes. Teachers have two possible responses: they can use the sails to harness the wind these other purposes create, and move them closer to maximising learning. Or they can turn on the motor to push the boat through the wind, getting the class to focus. For example, restating the purpose of the class may allow them to press on through despite the desire to maximise enjoyment, while making an exercise fun will use that competing purpose to help to maximise learning.

We spoke about the four types of learning environment: more learning happens in inclusive classrooms, because superstar students tend to dominate the learning in non-inclusive classrooms. We found that normalising silence can be a key tool in promoting inclusivity, as can participation software like Teachly. At the top of the pyramid, and what Dan aspires to, is 'team learning', where the students are not only *included* in class, but also take responsibility for their own learning. In a sense, the whole class is part of the teaching team.

We said that the learning environment can be defined as the bonds between the members of the classroom: the vertical bonds between the teacher and the students, and the horizontal bonds between the students. These form a lattice-like structure which can be strong or weak. The stronger the bonds, the more 'heat' they can withstand in the classroom. As a result, stronger bonds allow more cognitive work to be done, and more learning to take place.

Underpinning the idea of 'team learning' is the concept of psychological safety: the belief that one will not be punished for making a mistake. Google's researchers found was the single most important factor in building a successful team, and the same is true in a classroom.

Our first lecture back is the most challenging class of term for Dan to teach, and it does not go well. However, high psychological safety boosts *him*, as well as the students, and we will see how the strong learning environment helps him to manage the challenge.

PART III: INTERPRETING THE DATA

<h1>8</h1>

<h1>WHAT DO THE NUMBERS MEAN?</h1>

"This is a super-important class", says Dan at the start of Class 5. "It's like an airport class. We've done probability, and now we're moving onto statistical inference. The two are related, but you might not have a sense of that from your typical college course."

Dan is about to teach statistical inference, which is a fancy way of saying 'what do the numbers *mean*?'. Suppose we pick ten Americans at random and ask them whether they plan to vote Republican at the next election. Seven of them say 'yes', and three of them say 'no'. On one hand, perhaps 70% is higher than we were expecting: we heard someone say that the expected vote split was closer to 50-50. On the other hand, we might think to ourselves, "ten people is probably not a high enough sample size to give us much certainty about the wider population". The tools Dan plans to teach the class will help them to resolve this dilemma and tell them what they can infer, as well as what they cannot.

"I want to tell you a story about this", Dan continues. "In my PhD program, I learned about sampling distributions in an extremely theoretical way... I went through five years of graduate school doing lots of mathematical proofs, and realised later that I had no idea what a sampling distribution was. I think this class is very important to help you with that."

Whenever a new concept is introduced, there is some hidden tension created in the class. Dan has just used the phrase 'sampling distribution'. If you are reading and are not familiar with this idea: think about how your brain reacted to those two words on the page. You may have had an urge to switch off. You may be worried that we are going to spend the next few pages talking about sampling distributions, expecting you to understand. So far in Dan's class, we have been talking about probability and decision-making, which are part of everyday life. Sampling distributions are not. The brain has nothing to latch onto, which generates tension.

Dan responds to this tension in an unusual way. He shares his self-evaluations from previous years with the class, to give them a sense of the challenge he faces*.

---

* After every class, Dan writes reflections on what went well, and what did not. He keeps the self-evaluations of previous years and looks back on them before planning each class. The reflections for 2019 run to ninety-four pages, which also include feedback from students and course assistants. We will look at them in closer detail in Chapter 14. We spoke in Part II about Dan's obsession with gathering data about his class: feedback is a prime example.

"I'm going to go through a quick historical read of my own reflections over the last few years", he begins. "It started in 2011: 'this is a very important class and I did not like at all how it went.'"

The students start to laugh: Dan is showing vulnerability in a way most professors would not. He continues, reading his notes out loud:

- 2012: I struggled again in this class, and I cannot afford this: it is a very important class.
- 2013: This is a very important class. It went better than last year, but not great.
- 2014: The class did not go well again. Made a few changes that fixed some of the problems last year, but in the process created *other* problems.
- 2015: Okay, but not great.
- 2016: ('I like this one', he says.) For a while in the class, I had the illusion that things were going much better than in previous years, but quickly saw that it was just an illusion.

The students are laughing at each one. There is something reassuring about having a master of their field admitting that they do things badly too. Dan pauses, then continues.

- 2017: A total disaster. Need to rethink this class from the start.

This one gets a big laugh from the students: the worse it gets for Dan, the more the tension is released in the room. A lesson for *us* is that showing vulnerability can be an excellent tool for improving the invisible bonds in the classroom. He is bringing himself closer to the students. He concludes with last year's assessment:

- 2018: I think this class could have gone better but it's not hopeless. It's a very important class – keep trying.

This gets a laugh at the end and a few claps from the students. He has repeated the phrase 'this is a very important class' seven or eight times already. Of all the classes in term, he wants to be sure the students focus on this one: as the captain of the sailboat, he is using some of his fuel to push it on through towards its destination.

THE DICE GAME

Dan tells the students that *sampling fluctuations* are an airport idea: if we had taken a different sample, we would have got a different result. The 'sampling distribution' is everything we could have got: all the possible results from all the possible samples.

To illustrate this, he tries an experiment on the class. He places a six-sided die on each student's desk. ('Do not play with the dice yet!' he warns.) He explains that for their purposes today, the 'population' of interest is the 80 students in the classroom. This puts them in the rare position of being able to access the population: in real life, this is almost never the case. If you could speak to the whole population, you would not need to sample.

"You're each going to roll a die", he explains, "and if you get a 'one', you're going to be part of our first sample."

There is a murmur of anticipation. "It's okay if you're not selected into the sample!" he laughs. "I know you're all competitive. Alright: 1, 2, 3, go!"

Everyone rolls the dice. Dan explains that our goal is to use sampling to estimate the proportion of the students who are female. They know already that the room is comprised of 40 men and 40 women, so the true proportion is 50%. Of the 13 students who roll 'ones', 8 are female, an estimate of 62%. This is an overestimation of the true proportion.

A second roll of the dice finds seven 'ones', of which three are female: a proportion of 43%. This is an underestimation of the true value. A third roll gives us 6 out of 10: 60%, another overestimation. Dan is trying to embed in the students' minds that *different samples produce different estimates.*

"So if you ever work for a company, take a sample, and just give your boss a number, you haven't given them complete information. You haven't told them how much uncertainty surrounds that estimate. A sampling distribution allows us to measure that uncertainty. Does that make sense so far?"

There are a few murmurs of assent around the class, but mostly hesitant. Longer 'yes' noises which inflect upwards at the end are an indication of doubt, whereas shorter 'yep' noises that go down in tone are more likely to reflect certainty. At this point, the class is giving Dan more of the former than the latter. They get the thing with the dice about sampling fluctuations, but it's not clear how this 'sampling distribution' helps to measure the uncertainty.

Sensing this doubt, he encourages the class to "please ask all the questions you have". Psychological safety is important here: he does not want students to feel like their questions are delaying the rest of the class.

He gets five questions, one after another. Three of them highlight confusion about the sample size, which changed each time in the dice game depending on how many students rolled 'ones'. Dan confirms that the dice game was a pedagogical tool to get them to see the process of sampling, but in reality the sampling distribution based on samples of 10 ("everything that you can get when you pick 10 people") is different to the one based on samples of 13. Even as hands stop being raised for questions, there is an unease in the room at this new concept that many are yet to fully grasp.

He tries an animation on the board, which is designed to explain the concept with a 'toy' example in which samples of two are picked from a population of five, and the average age calculated. Once again, though, the students have lots of questions. For the first time this term, Dan is struggling to explain a new concept: there are lots of technical misunderstandings that are getting in the way. However, sharing his self-evaluations from previous years at the start of class was a clever way to keep the students on his side throughout this process. It is a difficult idea to teach.

THE FORTUNE 1000

One of the key ideas is that the greater the sample size, the more confidence you have that your result is close to the true one[*]. Dan highlights this by presenting a table showing the distribution of

---

[*] This depends on being able to sample randomly from the population. For example, if you are polling party members at the Democratic National Convention, you will not get a good estimate of how much of the wider US population is likely to vote for Trump whether you ask 10 or 1,000 people.

revenues from the Fortune 1000, the thousand biggest companies in the US:

Figure 8.1: Distribution of revenues in the Fortune 1000

Annual revenues (mean: $15 billion)	Percent of companies
Below $5 billion	48%
$5-10 billion	22%
$10-25 billion	18%
Above $25 billion*	12%
Total	100%

If someone asks, "what is the probability that a Fortune 1000 company selected at random has annual revenues above $25 billion", we can easily read from the table that the answer is 12%. The question Dan asks on Poll Everywhere is slightly different:

"If we take a random sample of 100 companies from the Fortune 1000, what is the probability that the average revenue in this sample will be greater than $25 billion?"

A. Less than 12%
B. Equal to 12%
C. Greater than 12%
D. Not enough information
E. Don't know

Dan looks privately at the results, which show that most of the class does not have the right answer. He asks the students to form groups of 2 or 3 people, ideally with different answers, and to try to convince

---

* The highest, at $514 billion in revenues, was Walmart, followed by Exxon Mobil and Apple.

each other of their answers. The key insight is that the more you increase your sample size, the harder it is to get an average as high as $25 billion. You might get lucky, but as the sample size increases, the chances decline that you just keep picking big companies. Instead, you get more likely to be closer to the mean of $15 billion. So the probability that the average revenue will be above $25 billion is *less than* 12%, even though 12% of companies have revenues above that number.

After the 'convincing' period, Dan runs the poll again. The number of right answers increases, but 4 in 10 students still believe the probability will be equal to, or greater than, 12%. For Dan, this is a disappointing result: there is likely to be something he can change about the class to improve the understanding next year. He calls on John, the American student, who he knows is likely to have the right answer: this is probably a time to lower the heat and provide clarity, rather than orchestrating conflict. He also has one eye on the time, as all the questions earlier mean he is running behind schedule.

John points out that if the sample was just one company, then the answer would be exactly 12%. If your sample was all 1,000 companies, then you know for sure that the mean will be $15 billion, so the probability of being above $25 billion is zero. So as you go from a sample of 1 to a sample of 1,000, the probability must fall from 12% to zero. A sample of 100 must therefore give an answer of 'less than 12%'.

Dan spots an important learning point and intervenes: "I would like people to respond to John, but I do want to say one thing about his reasoning that is helpful for all of us to learn from. When you get a difficult problem, it's often helpful to do what he just did and think about the extremes."

Others chime in to discuss their approaches to the problem: one student mentions how his group initially had three different answers, but are now united around one, suggesting some progress within the group. But the discussion is not quite as rich as if he had called on a weaker student than John to start with: the time constraint meant this was something he had to concede.

How sampling makes everything normal

Dan closes out the class by talking about the effect of sampling on outliers: "I want you to think about those very few companies that earn way above the others – Apple, Walmart, and so on. Imagine all the possible samples that exist, and imagine Apple being included in many of those samples. Apple will pull the average up, but the average of the sample *relative* to Apple will be lower. *Sampling pulls the outliers in.*"

The distribution of the revenues of Fortune 1000 companies is highly skewed to smaller numbers: there are lots of companies with revenues below $5 billion, and very few with revenues above $25 billion. This is different to the distribution, say, of women's heights: the most common height is about 165 centimetres, and heights get less common as you go further away from that number in either direction. Most women are between 155-175cm tall. We call the distribution of heights a 'normal' distribution because there are so many things in life that look like this: heights, weights, and IQ scores are all 'normally distributed'. When you are almost certain of something, the bell curve will be tall and thin: there is only a small range of highly likely values. When there is high uncertainty, it will be short and wide.

Figure 8.2: two common types of distribution

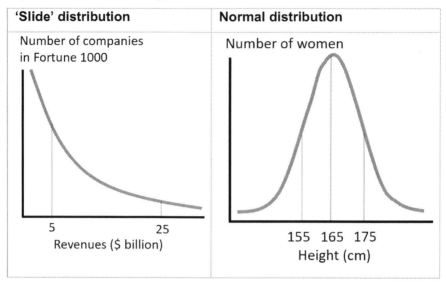

Something remarkable happens when you take samples: as if by magic, the 'slide' distribution on the left turns into the 'normal' distribution on the right. If we look at the distribution of the *average* across all 100 companies in the sample, it is a normal distribution. If we keep taking random samples, we can expect most of the averages of those samples to be between $14.9 million and $15.1 million, very close to the mean of $15 million.

"You should be getting goosebumps!" says Dan. For mathematicians, the transformation into normal distributions by sampling is a magical result called the Central Limit Theorem that generates huge excitement. In statistics courses, this usually leads the teacher to waste everyone's time by proving it mathematically. To the relief of the students, Dan does not.

THE POST-MORTEM

Dan stays behind as always to answer students' questions, but in the post-class meeting with the teaching assistants, he is visibly frustrated. Once again, he feels he struggled to get the key ideas across to the students. Dan is a 4.9-rated professor at Harvard: if it happens to him, it happens to everyone. His timings were also off: rather than taking 15 and 30 minutes respectively, the first and second sections took 24 and 45 minutes, meaning he had to miss out a section explaining the change in the distributions.

We discuss the potential areas for improvement. There are pedagogical trade-offs in the dice game. The key benefit is that it shows the act of sampling; it is also fun, and another experiment performed on the students, giving them responsibility for their learning and strengthening the bonds in the classroom. The downside is that it creates an early opportunity for confusion as to the nature of the sampling distribution. As some of the questions alluded to, there is a different sampling distribution for each sample size: the samples you can 'get' by picking 10 students at random are different to the ones you can get by picking 13. The dice game will lead to varying numbers of students picked, and people could be forgiven for wrongly concluding that there is just one sampling distribution which includes samples of all sizes.

We wonder whether the toy example, in which an animation shows the process of all possible samples of size two being taken from a population of size five, might also create confusion. In the real world, populations are big: that's why we need to sample. Does the conceptual benefit of showing a sampling distribution built from scratch outweigh the cognitive dissonance the students might face in thinking of a population as a small thing? And did it lead to awkward questions about whether 'replacement' is allowed – that is, including the same person twice in the same sample?

One of the course assistants suggests that the leap from the toy examples to the Fortune 1000 question is bigger than Dan thinks it is, and that having three separate contexts (gender, age, and revenues) is too many. Dan writes in his reflections: "I think she is right". Another course assistant suggests that they be asked to do more preparation work before class on this subject.

We wonder about possible improvements. Making clear that the sampling distribution depends on the sample size N is one thing: perhaps we could show an animation that starts with $N = 1$, and increases N all the way up to $N = 1,000$. We would see the distribution turn, as if by magic, from the original 'slide' to a bell shape, and that bell shape would get thinner and thinner as we get more and more certain of being close to the true average.

Dan also gets a couple of emails from students in the coming days, keen to support. One enjoyed the first part of the class, but was then confused by the Fortune 1000 question. Another felt that the Fortune 1000 question was the most important bit, and that Dan spent too much time early in the class with toy examples rather than real-life ones. He writes that "seeing concepts in action will help students to grasp them". This is an indication of the challenge Dan faces: students will learn in different ways, and he wants to give them the best chance.

For those interested in pedagogy, there is little better than throwing these ideas around with Dan after a class. Everything is a trade-off, and there are undoubtedly teachers out there who have mastered these trade-offs better than him. However, his drive to improve, and the obsession with the best possible way to teach a concept, comes across loud and clear to his students. The knowledge that Dan is working so hard to teach them strengthens the contract: the students start to think, "he is taking this seriously, and so should I".

After all the discussions are over, Dan writes his reflection for this class, just as he has done for a decade. At the top is the one-line summary:

- 2019: After having made some progress on this important class last year, this was a setback. I need to rethink it once again.

9

ANSWERING QUESTIONS WITH STATISTICS

To try to make sure no-one is left behind after struggling with Class 5, Dan adds a question to the next problem set. The students are asked to work in groups to produce a slide explaining sampling distributions to a 'smart policymaker not well-versed in statistics'. The best group will present it in a class next week. This is another key example of Dan getting the students to take responsibility for their own learning: they are being asked to *teach* the idea, not just to explain it. In teaching someone else, they are forced to think pedagogically about how they can make this complex idea simple.

In the real world, we are never going to have access to the wider population. We start off completely in the dark: statistics is a field predicated on uncertainty. The best we can do is to sample a few of the population: in a poll about US politics, we might take a sample of 1,000 people. With the US population around 300 million, we have polled just 1 in 300,000 of them. How on earth can we be expected to make judgments about the whole population based on the opinions of such a small fraction?

This is why knowing that sampling distributions are bell-shaped is so important for statisticians. There were a huge number of possible samples, and we only saw one of them. We do not know how ours ranks in relation to the others. But we do know that the distribution of all those possible samples is bell-shaped, and we know that the higher the sample size, the thinner the bell curve is. If it is thin enough, then we can be confident the true population value, which we never see, is close to the one in our sample. We would be unlucky if it is not.

It turns out that the size of the population is not important: our level of certainty is determined by the size of the sample and the extent to which the population varies on average. Class 6 deals with 'confidence intervals', which measure the uncertainty. Dan puts up a recent news clipping[xxvii], which says[*]:

"A recent NBC poll indicated that 31% of Democratic Primary voters favour former Vice President Joe Biden. The poll also indicates support for Elizabeth Warren (25%) and Bernie Sanders (14%). The poll was based on a sample consisting of 506 Democratic Primary voters."

He asks them to calculate the margin of error for the poll. A rough guide is that to be 95% sure that the margin of error includes the true value, you need to take $1/\mathrm{sqrt}(n)$ either side of your estimate, where n is the sample size. The eagle-eyed reader may spot that the higher n is,

---

[*] This class took place in September 2019, before Joe Biden won the Democratic nomination for the 2020 Presidential elections.

the smaller the margin of error. The class gets out their calculators, and they find that $1/\sqrt{506} = 4.4\%$. One of them gives Dan the answer.

Dan moves to the next slide, which is the footnote of the NBC poll:

"The NBC poll was conducted Sept 13-16 of 506 Democratic Primary voters, and it has a margin of error here of plus-minus 4.4 percentage points."

He points out that the numbers are the same: "Magic! You just calculated the same margin of error that NBC reported." Moments like this help to build the confidence of the students: they see their own calculations affirmed. The fact that this is a real-life example means they can trust immediately that they will be able to recreate the calculation in the professional world.

The class then finds the confidence interval: with a margin of error of about 4 percentage points and a sample estimate of 31%, they can be 95% sure that the true support for Biden is somewhere between 27% and 35%. The wider lesson is one about their relationship with statistics: point estimates always have uncertainty around them. Often, the size of the confidence interval is more revealing than the estimate itself. This will be an important theme in Class 7, which introduces the idea of hypothesis testing.

HYPOTHESIS TESTING

"Who in the classroom has kids?" asks Dan, at the start of Class 7. Several hands go up, including that of Ankit, an Indian student who has a young daughter. "I want to introduce to you a little girl", Dan says. He flicks to a picture of a small girl on the screen, holding a toothbrush and looking mischievous. The class laughs.

"Here is the scenario", Dan continues. "Suppose Ankit asks his little girl, 'did you brush your teeth?', and she says 'yes'. We're going to assume that she's truthful, and that she *did* brush her teeth. That's our null hypothesis. Now, let's say Ankit goes to the bathroom, puts his finger on the toothbrush, and finds that it's dry."

He pauses, and the class laughs again. "What should he conclude?" There is a silence. "Let's start with this: would it be *impossible* for him to find a dry toothbrush if she had brushed her teeth?"

There are a few 'no's murmured from the class. "Can anyone give me ways it might have happened?" asks Dan.

Mila, from Kenya, has a suggestion: "she might have brushed her teeth with *his* toothbrush". The class laughs, and Dan adds, "Your smile makes it seem like you were in the same situation once, but I'm not going to press you on that". The laughter grows.

Hiro, from Japan, suggests that they might be living in a very dry climate, meaning that in the short time after brushing her teeth, the toothbrush had already dried out. Another student says she might have an extra toothbrush. A third suggests she dried it herself with a hairdryer. The students laugh each time, strengthening horizontal bonds in the classroom.

"Okay", says Dan, "you've all come up with possible explanations for why the toothbrush might be dry even if she *did* brush her teeth: that is, even if the null hypothesis is true. We need to ask: how unlikely is this to happen?"

The class suggests that the probability is low: the evidence does not look good for the little girl.

Dan continues: "So Ankit is going to go to his daughter and say, 'I reject your claim.'"

More laughs. "Maybe not quite like that", Dan says, "but the important thing is that he is rejecting the null hypothesis not because it is *impossible*, but because it's very *unlikely*.

Given the null hypothesis is true, how unlikely is it that we get a sample like the one we got? That's called the *p-value*. The p-value is at the centre of hypothesis testing."

If you are ever unfortunate enough to be sat next to a researcher at a dinner party, they will probably talk about p-values a lot. Scientists and economists spend their lives in search of low p-values. When your p-value is low, it says "it's very unlikely that we would have got these results by chance, so they must be reflecting a real finding". A popular meme shows '$p = 0.051$' as one of the scariest things an economist can see at Halloween, because the threshold for significance is often set at $p < 0.05$.

In this case, the class is saying that Ankit's p-value is less than 0.05: it's possible his daughter brushed her teeth, but highly unlikely given the dry toothbrush.

Dan explains to the class why he's spent so long on the toothbrush story rather than just telling them the idea up front. "It's very easy to get lost in the sampling distributions and the p-values and miss the important logic. Whenever you do a hypothesis test, this is what you're doing."

He pauses and invites questions. There are only a couple of clarifications: it seems like the class gets the idea. Having introduced the concept, he now wants the students to take over.

He moves on: "Here's a real-world example, where one of us was involved". The screen shows a study about women's empowerment in Nicaragua, along with the photo of a Peruvian student named Andrea, who co-authored the study. As it happens, Andrea's father is in the audience today: Dan saw an opportunity to include her in class and make him feel proud of his daughter*. He asks Andrea to describe the study to the class.

Andrea explains that they were helping Nicaraguan women to buy livestock, reducing food insecurity in rural areas by providing a source of meat, eggs, and milk. They found that this reduced women's disempowerment by 7% based on their index, as women exposed to the program became more involved in their communities and participated in group training.

Dan asks, mischievously: "Was the study based on every woman farmer in Nicaragua, or did you take a sample?"

Andrea smiles and replies: "It was a sample". The class laughs.

Dan reaffirms the learning from earlier: "Because it was a sample, we have uncertainty, so we have to do hypothesis testing. In her case, the

---

* Lots of parents find their way into Dan's class when they visit the Kennedy School. It is no coincidence that students feel comfortable bringing their parents into this class: it reflects the high psychological safety of the environment, and the interest Dan shows in their backgrounds.

disempowerment score was reduced by 7%. Should we conclude that the program reduced disempowerment by 7%?"

There is a silence as students arrange the information in their heads. Eventually, one student suggests that it depends on the p-value. This is not quite right, but Dan does not invalidate his response. He moves onto another student, who correctly says that we cannot conclude the true impact was 7%. We *can* have some confidence that the true number is in some interval around that, but we cannot be certain.

Dan confirms that this is right: "7% is our best guess, but because of sampling fluctuations, it's likely to differ from the true impact in the population. In the study, the confidence interval was between 4% and 10%, so their 'impact' is likely to be somewhere between those two figures."

Having worked through Andrea's example, and having got the class to give her a round of applause as she sat down, he gets the students to tackle a question:

"An analyst claims that the 20% of the households in your country are poor. You are doubtful and draw a random sample of 900 households, and observe that the proportion of poor households in your sample is 23%. Should you conclude the analyst is wrong?"

This is a classic example of the sort of question that statistics can help you to answer. Dan clarifies how he wants them to go about it.

"You see, mathematically, that 23% is different to 20%. The question is whether they are *statistically* different. Is 23% so different from 20% that we're going to reject the analyst's claim? We're going to see how inconsistent our sample is with that claim. How likely is it, assuming the proportion of poor households really *is* 20%, that we see a result that's as much as 3 percentage points away?"

Now, Dan decides to sequence the flow of information by offering up a hint. He could just ask the class to run the hypothesis test, but he knows that many will run into mistakes or feel apprehensive about what to do. He wants to nudge them onto the right path by ensuring they at least set up the test correctly.

An alternative could have been to let them make those mistakes, and then learn from them. His experience tells him that this would not be productive: if the students start off on the wrong track *here* then it is difficult for them to recover, and could potentially derail the class. Instead, he will show them together the mistakes that he hopes they will avoid.

He pauses, and asks the students: "What do you think the null hypothesis should be here?"

Chris, the Australian student, correctly answers that the null hypothesis is that the true proportion of poor households in the population is 20%.

Dan pushes: "That's good, but can you to explain to me: why didn't we test whether the true proportion is 23%?"

Chris correctly answers that the *claim* is that it's 20%, and this is what we base the null hypothesis on. Ideally, you make your hypothesis before knowing anything about the sample.

Dan asks another question: "Why didn't we test whether the *sample* proportion is 20%?"

Chris says he is not sure. Another student chimes in: "We already know the sample proportion is 23%. We're testing the *claim*, not what we got in the sample."

Dan is being pedantic on this point because it is a common source of misunderstanding. When these concepts are new, it is easy to mistake the 'true proportion' for the 'sample proportion' and to set up the wrong hypothesis test as a result. Having written the three possible hypothesis tests on the screen ('true proportion is 20%', 'true proportion is 23%', 'sample proportion is 20%'), he crosses out the second and third, and writes 'NO!' in capital letters, getting a laugh from the class. He then gives the students some time to do the hypothesis test.

After a few minutes, he asks the class to discuss what they did with a neighbour, then brings them back together. He gets students to walk him through the calculation, and repeats the importance of setting up the test correctly:

"Here's what I want you to remember five years from now, when you've forgotten everything else. When someone's doing a hypothesis test, I want the voice in your head to be saying *assume the null hypothesis is true... assume the null hypothesis is true*."

The calculation shows that the p-value – the likelihood that they would have seen an answer as big as 23% if the true proportion were only 20% – is just 0.02. That suggests that to see 23% in the sample would be very unlikely, and so they reject the claim[*] that the true poverty rate among households was only 20%. Just as we rejected the girl's claim that she

---

[*] If the p-value had been above the threshold for significance of 0.05, then they would not have had enough evidence to reject the claim.

brushed her teeth because her toothbrush was dry, we are rejecting the analyst's claim that the true poverty rate is 20% because our sample rate of 23% was too far away.

The students are moving into the 'nuts and bolts' of Statistics. Conducting hypothesis tests is a mechanical process, and the temptation is high for a teacher to go through the mechanics without a deep understanding of the logic. Dan's goal is to focus first on the logic, and then teach the mechanics. In doing so, he promotes understanding over memorization.

MID-COURSE FEEDBACK: RESISTING STUDENT DEMANDS

Just as he encourages the students to do, Dan uses statistics to answer many of his own questions about the course. As we have already seen, his collection of data is integral to the way he designs the course and adapts to student feedback. He also wants to share this data with the students themselves, as this helps to put responsibility back on them for improving their own learning experience.

Since the course runs from the start of September until mid-December, and we are still in September at this point, the term 'mid-course feedback' is misleading. Dan wants to gather the data as early as possible in the class so that he has the maximum possible time to adapt. If there are things he is doing wrong, then waiting for after the midterm exams to fix them is probably too late. One trick he uses for collecting the feedback is to make it a part of a problem set, so that the students *have* to respond.

"First of all", he says, "I want to say thank you for doing the feedback survey. I do realise that it was part of the problem set, but the response rate was spectacularly high."

The class laughs at the fact that they had no choice in the matter. "I want to spend ten minutes telling you what I heard, and what I plan to do in response to your feedback, and what I *don't* plan to do in response."

There is another chuckle from the class in anticipation of Dan holding his ground. He puts up the summary statistics for 'pace' and 'difficulty' based on the student responses:

Figure 9.1: student responses when asked about the pace and difficulty of the course so far

Pace		Difficulty	
Answer	Percentage	Answer	Percentage
1 – Very Slow	3%	1 – Very easy	1%
2 – Slow	12%	2 – Easy	16%
3 – Adequate	73%	3 – Adequate	69%
4 – Fast	11%	4 – Difficult	14%
5 – Too fast	1%	5 – Too difficult	0%
TOTAL	100%	TOTAL	100%
Average: 2.96		*Average: 2.94*	

Dan continues: "In terms of the pace and difficulty of the class, most of you think it's adequate, and approximately the same percentage thinks it's slow or fast. Given the differences in the class, the 2.96 average feels close to optimal. I do realise that those of you at the extremes are not going to be satisfied: if that's you, please come and talk to me to see if we can do something."

He then makes the point that the class is a very heterogeneous one. "Here is the number of statistics or econometrics courses that you took before coming here", he says, putting another table on the screen. "Roughly a third of the class have two courses or fewer, then there are a few of you at the top end who have taken more stats courses than I have."

This gets a big laugh from the class: he is bringing himself down to their level, and putting the responsibility for learning back on them.

Figure 9.2: statistics or econometrics classes taken prior to this one

Courses	Percentage
0	1%
1	10%
2	27%
3	21%
4	28%
5	6%
6 or more	6%
TOTAL	100%

"This is the level of satisfaction with the instructor", he continues. "In general, the class seems to be very satisfied, though there's certainly room for improvement. Those of you who are unsatisfied – please do come to my office hours, and we'll see if we can do something."

Figure 9.3: class ratings for 'satisfaction with the instructor'

Answer	Percentage
1 – Very unsatisfied	3%
2 – Unsatisfied	1%
3 – Neither	0%
4 – Satisfied	12%
5 – Very satisfied	84%
TOTAL	100%
Average	*4.73*

The average of 4.73 is very high in the context of Harvard ratings, but Dan's aim is to improve it to his high bar of 4.90 and above by the end of term. A crucial part of that is identifying early the problems people are having with the class, and helping those people. A hint to those problems is found in the qualitative section of the feedback, where students are given a chance to voice any concerns or praise. Although students loved the real-world examples and the teaching style, some have concerns about the problem sets.

"The problem sets received mixed reviews", he begins. "The main complaint is that although 92% of you find them helpful, they are *very* time-consuming."

He puts a quote on the screen from an anonymous student. "I feel like the length of the problem sets yields a mentality to simply 'get it done', rather than reflect and process what we are doing", the student complains.

"I like the indirect way you are telling me, 'you are wasting our time!'", Dan replies with a smile, generating a big laugh from the class. "There was a cathartic element to this survey. I do want to put things in perspective. The average time spent on problem sets is 9 hours, and the median is 8 hours. At the beginning of term, I told you they should take roughly 8 to 10 hours. So the mean and the median are okay: the problem is that there's a very wide distribution. 12% of you are spending 15 hours or more on each problem set: that's too much."

He pauses. "I need to understand better where those hours are going, and I'd like those 12% of you to come to my office hours and help me with that. In the next problem set I'll also do a survey asking you to measure time on each question, and hopefully things can get better.

"Those of you who are in that 8 to 10 hour bracket: I think the only way you'll learn statistics is by doing, and what we do in class is just one component of that. But if you're spending more than 15 hours, that sounds like it's running your life, and I know you have other courses to do. I want you to have an experience that's not just about this course.

"So in response to your feedback", he concludes, "we will scrutinise the problem sets to make sure they're delivering learning efficiently, and we'll try to post them earlier to give you more time. It would be easy for me to make them shorter, and I'm sure that would increase your satisfaction. That's not what I see as my goal: my goal is to maximise your learning, and I hope you guys respect that. Learning and satisfaction are positively correlated, but if I had to choose one, I want you telling me five years from now that you know this stuff deep into your bones, rather than telling me you were very happy with the class. I want to emphasise: I'm here to help you learn, so please come to office hours if I can help in any other way."

As well as offering support for those students spending too long on the course, Dan is trying to focus the rest of the group towards further work in future problem sets. In the metaphor of the sailboat, he is again using some of the fuel to push the boat through the headwinds. There is only a finite supply, and he can only push the students so far. But a failure to use the motor at all would mean being pushed off course, away from his goal of learning maximisation.

As well as understanding the students' experience, Dan also wants to know which bits of the course are sticking in their minds, and what they are finding helpful. In the next problem set, he includes the following question:

PART 2A – NERDY FRIEND

Suppose a friend from college calls you and asks how this course is going. This friend has <u>not</u> taken a statistics course before. After giving him your personal opinion on the instructor, his teaching techniques, and how lengthy his problem sets are (yes, I know what you are thinking right now), your friend asks you to summarize the key concepts you have studied so far in this course.

You decide to write him an email (about 2-4 paragraphs long) explaining the key concepts we have studied so far in the course, and how they relate to each other. What would you write to your friend? Make sure you include the concepts of probability, sampling distributions, confidence intervals, and hypothesis testing.

As we will see in the end-of-course feedback, this question divides opinion among students, some of whom do not feel like being dragged back through all the class material in a problem set that is already long. However, Dan believes that getting them to explain the concepts to a friend not well-versed in statistics is an important way of reinforcing the material early in class. The topics that follow in the next few classes will rely on a solid understanding of these concepts.

CHERRY-PICKING: THE 2009 ELECTION IN IRAN

The protests on the streets of Tehran, fuelled by many years of anger and resentment, had turned violent overnight. Buses were overturned and set alight, sending up rising smoke that could be seen for miles. The anger grew into the Green Movement, in which millions of Iranians marched against the regime. Some shouted "death to the dictator", others "give us our votes back". They were angry that Mahmoud Ahmedinejad, the hard-line incumbent, had won re-election, and believed he had manipulated the vote. His support in rural areas had mysteriously jumped, allowing him to win far more votes than expected.

The first question in Dan's fourth problem set takes us back to 2009, and considers the allegation of fraud, based on the vote counts across Iran's 116 provinces. Looking just at the final digit of the vote count for Ahmedinejad (e.g. if the vote count for a province is 5,678 then the final digit is 8), a suspicious anomaly was found. If the last digit was virtually random, as we would expect in the event of no fraud, we would expect roughly 10% of the vote counts to have final digit 0, and the same for all the other final digits from 1 to 9.

Figure 9.4: distribution of last digits of vote count in Iran

Last Digit of Vote Count	Number of Provinces	Observed Percentage
1	11	9.5%
2	8	6.9%
3	9	7.8%
4	10	8.6%
5	5	4.3%
6	14	12.1%
7	20	17.2%
8	17	14.7%
9	13	11.2%
0	9	7.8%
Total	116	100%

However, instead we see that 17% of the provinces reported a vote count ending in '7', well above the 10% expected. The question is whether this was due to sampling fluctuations, or whether there was foul play involved. As part of the problem set before class, the students are asked to test the hypothesis that the underlying likelihood of vote counts ending in '7' is equal to 10%.

The first line of their answer has been drilled into them: "assuming the null hypothesis is true". This means they are assuming the underlying likelihood is 10%. They want to know how unlikely it is that as many as 17% of the vote counts will end in a '7'. This is the p-value, and they calculate it to be just 0.01: less than the traditional value of 0.05 for significance. The next question asks them to make an inference:

"Would you conclude that there was fraud in the 2009 Iranian elections?"

Dan gets them to answer this in an online form so that he can access all their answers quickly. He is bridging the gap between theory and practice again here: a hypothesis test is useless if you cannot make valid inferences from its results.

Before he brings up the question in class, he asks the students to write down a random number between 10 and 99. Having done so, he gets them to 'vote' on Poll Everywhere for what the last digit was. "If you picked 53, the last digit is '3'. Everyone clear?" He smiles. "You guys are like, 'we made it to Harvard, and this guy's asking us if we can think of a number.'"

There are laughs from the class, releasing tension. Before this section, Dan had run through some technical material on different types of hypothesis test, which some students may have had trouble following. Small exercises like this, which require no cognitive work but which will contribute to the experiment, give the students a break and allow their brains to get ready for the discussion to come.

Dan continues: "If we are able, collectively, to actually pick random numbers, what do you think the distribution should be?" The class murmurs words to the effect that it should be a uniform distribution in which each last digit has probability 10%.

"Shall we look at the results? I'm kind of nervous right now", he laughs. Dan is building anticipation: even for something as inconsequential as what random numbers the students picked, this kind of pause gets them on the edge of their seats. This is a good example of the idea that he does not need to come up with an elaborate routine to be entertaining. The simple act of withholding information for a few seconds is enough to build suspense, which has a payoff at the end for both the students and the teacher in the form of an 'aha' moment.

The results show a varied distribution, with 23% of the class picking a number whose final digit is '7'. There is a big 'ohhh' from the group. One of the students, remembering the problem set, shouts out "fraud!", which gets everyone laughing. Dan acknowledges to the students that '7' was the most frequent number picked, and tells them: "What you have done is consistent with a lot of evidence that human beings are bad at making up numbers."

He goes on to explain: "We tend to pick random-looking numbers like 37; we are bad at picking numbers that don't look random, like 50; we tend to pick adjacent final digits too much, and we don't repeat digits often enough."

At first glance, this might seem like nothing more than a quirk of the human brain. But the connection with election fraud piques the students' interest. The idea that we can use this to investigate wrongdoing has something genius about it.

Dan asks: "In the problem set, I asked you to look at this, focusing on the last digit '7'. You tested whether 17% was different to the 10% expected, and found a p-value lower than 0.05. The question was: is this evidence of fraud?"

Rajiv thinks so: "Because of the low p-value, we have evidence that it's not random. If we believe the only way it's not random is because of fraud, then there's evidence of fraud."

Juan disagrees: "I don't think this is a valid test. We've just selected the digit with the highest proportion. We should have tested the likelihood of *at least one* of the 10 digits having a value as big as 17%."

Oliver continues Juan's line of thought: "Yeah, there's ten chances for us to get a fluctuation, so the odds of having *one* of them having a p-value of below 0.05 become much more likely."

Fernanda chimes in: "I was thinking you could simulate this and always look at the 'worst' one – it wouldn't be that unlikely for that to give a significant difference from 10%. I think it'd happen quite a lot."

Dan explains to them that they have just described the idea of 'cherry picking'. This is another airport idea that he wants them to be able to describe to him in five years' time, and on the side screens is a colourful image of a girl picking bright red cherries. "As a researcher, you don't get to choose the test that most favours your result. In this case, we saw that '7' was the most common number, and then *chose* to do the test on the proportion of '7's. This is cherry-picking."

Dan wants to make sure the class spends some time on this analogy, and is also keen that students for whom English is not a first language do not lose out from the use of an English idiom. "Does anyone in the class have experience picking cherries? Why is this a valid comparison to make?"

Emma, an American, has been cherry-picking several times. "When you go cherry-picking, there's lots of different cherries out there but they're not all made the same. You're usually charged by weight, so you want to pick the cherries that are tastiest. So you're just picking a subset of everything in the population."

Dan nods. "This is what I asked you to do in the problem set: I cherry picked the test you were going to conduct. This is a problem that permeates a lot of the research you see, and a problem across the world. I wonder if you can give me examples of how you might have been the victim, or the perpetrator, of cherry picking before you came to Harvard?"

A student coming from the financial sector points out that some stock indices are constructed to include the biggest companies. The ones that fail, rather than showing up as making a big loss, aren't included in the index anymore and are replaced by others. This survivorship bias is a form of cherry-picking.

Monique, a consultant, admits that she has cherry-picked when coming up with 'market comparables' for a company: these are similar companies in the market used as benchmarks when coming up with valuations. She says, "You don't just pick a random subset. You pick one that makes your valuation look attractive."

Dan takes a few more student comments: this is a popular topic, with lots of the class having witnessed some form of cherry-picking in their professional lives. He then looks at Gabriela, who he knows has worked in academic research, and asks her if she's ever seen cherry-picking. She thinks out loud: "With papers, when you don't find a statistically significant result then you don't publish anything. You always see papers with significant results, but we never see papers published with no significant result."

Dan confirms: "It's much harder to publish a paper with no significant result, as I'm currently discovering with this study I've just finished." This gets laughs of sympathy from the class.

He wraps up: "These are all great examples. I wanted to spend a few minutes on them because I want you to see how widespread this practice is in the world. I hope that you will not do cherry-picking yourselves, or at least recognise when you're being asked to do so. If we are looking to build a base of knowledge to make better decisions, this damages those objectives.

"When you conduct a study, or read the literature, you should be very aware of the potential to be selective in deciding what results to present, or what tests to conduct. If you see results that seem different to what the initial hypothesis set out to look at, you should be highly suspicious of cherry-picking."

As Dan suggests, the problem of cherry-picking is widespread. It is not an exaggeration to say that it is an existential threat to some fields of academic research, and an underappreciated flaw in the hypothesis testing process. What if the papers that find significant results just happen to be the 5% of experiments that we would *expect* to find significant results purely by chance? Perhaps the other 95% are simply not published. One way to test this is to try to recreate experiments that have led to significant results in the past. The results so far have not been encouraging, with replication rates as low as 1 in 3 in some fields. It is possible that two-thirds of results just came about by chance.

This has been termed the 'replication crisis', and is yet to be resolved. In fact, the replication crisis is just another form of the prosecutor's fallacy – the logical flaw that condemned Sally Clark. Remember that this came about when the prosecutor reversed the conditionality of two events, and looked at the likelihood of two cot deaths given that she was innocent, instead of the likelihood of her innocence given two cot deaths. The second was much more likely than the first. In hypothesis testing, we do the same thing: we *assume* our treatment has no effect, then consider how unlikely our results are if that assumption is true. What we actually want is the likelihood of our treatment *having* an effect, given the results we obtained. Perhaps some of these 'successful' treatments just happened to get lucky.

So were the 2009 elections in Iran subject to fraud, or not? Cherry-picking is another example of the art of interpreting statistics being more important than the science of generating them. Dan gets the students to run a different test to correct for the first cherry-picking problem: this test looks at the whole distribution of last digits in the data and tells us how unlikely that distribution would be if the true likelihood was 10% for each digit. They get a p-value of 0.08: still quite low, but not below the threshold of 0.05 to be significant. Under *this* test, we would not be able to say confidently enough that there was fraud, because we would get results like that 8% of the time even if there was no fraud.

Additionally, we still have the wider cherry-picking problem to contend with: while it might be tempting to say that we are "92% confident of fraud", to do so would be to fall prey to the prosecutor's fallacy. Perhaps it just happens to be in the 8% of fair elections across the world that we *expect* to look like this. What if we had been looking at the results every time an election happened, just waiting for something like this to show up? Looking at it this way, it seems like we cannot have enough evidence to conclude that electoral fraud took place.

However, the argument swings back and forth. Maybe we were not cherry-picking at all when we decided to investigate the number 7. As we saw in the random numbers generated by the students, and in the academic literature, 7 is the last digit we *expect* to see most often when humans try to make up numbers. It depends on our intentions before we saw the data. If we started by saying "let's see how many 7s there are", and *then* found lots of 7s, then this is not cherry-picking. If we had no prior idea about which numbers would come up the most in the event of fraud, then to see the data and run a test on the number 7 *is* cherry-picking.

The difficulty is that we will never know. If you wanted to discredit the government, you might act like you were looking for 7s all along, even if you only decided to play it that way *after* seeing the data. Similarly, researchers might run 20 tests, see a significant result in one of them, and then rewrite the paper to ignore the other 19 and claim that that one result was what they were looking for all along. Dan's advice to the students is to be wary of this kind of manipulation, and when conducting studies, to outline at the very beginning what you are looking for.

I think there probably *was* a degree of fraud in the elections in Iran. But that's a personal opinion: a conclusion based on my prior beliefs (he's the sort of candidate I would not be surprised to see manipulate an election) and the strength of the evidence we have amassed, even if it leaves room for reasonable doubt. The best we can do is to 'be a Bayesian' and use the results of the test to update our opinions. The underlying point is that even after running the numbers, we still rely on our own judgment: statistics is not mathematics.

TRANSLATING TO THE REAL WORLD

The class has learned how to answer questions with statistics using hypothesis testing, and how to avoid pitfalls like cherry-picking when doing so. The wrap-up class comes the day before the students' midterm exam, and is deliberately light on statistical content. However, it serves as a nice end to this part of the course. Having immersed the students in hypothesis testing for the last few weeks, which at times can get quite technical, Dan brings them back up to the surface with a class on interpretation.

The class starts with a presentation from Henry, a Canadian student. Recall that after the difficulties of Class 5, Dan added a question to the next problem set asking the students to get into groups and come up with a slide to explain the concept of sampling distributions. Henry's group was selected for having the best slide, and he now explains sampling distributions to the class. He shows multiple samples of size 100 of human heights on the screen, and considers the mean of each one, before showing the distribution of those means on the next slide. As predicted by the Central Limit Theorem, the sampling distribution is normal.

"Magic!", says Henry, getting a laugh from the class for imitating Dan.

Once again, Dan has put the responsibility of learning a difficult topic on the students themselves, and they have delivered. As the midterm exams will show, the level of understanding of sampling distributions in the room is much greater than it was in Class 5. Some might consider this a failure on Dan's part: perhaps his lecture was not clear enough, and this meant the students had to go off and learn it by themselves. But the humility to accept that students will not always understand a tricky concept right away, coupled with the knowledge that they will understand it better if they have to explain it to someone, is a central part of his teaching approach. In this case, it appears to have produced positive results.

This humility is closely linked to psychological safety. If the students themselves are humble enough to know they may get things wrong, or even be wrong about things they know, then being wrong is no longer a source of shame. One student, asked about his biggest takeaways from the course, talks about his "increased humility, gained from the many class discussions where I found out I was wrong about an idea". Humility is a skill that can be taught and practised, and students leave the course feeling more ready to be wrong in other aspects of life. This is likely to serve them well, and to help them to create psychologically safe spaces in the real world when it is they, not Dan, who are responsible for them.

After Henry, Dan presents an experiment in which farmers in Vietnam were given training, and ended up earning an average of $100 more as a result. He asks the class to remind him what key phrases like "this effect is statistically significant at the 1% level" mean. The students have gone through these questions before, but the intent is to get those students who knew nothing about hypothesis testing before the course to answer them.

Dan waits for hands, commenting on progress ("and many hands are going up here... many *more* hands are going up") to draw more students into raising their hands. Eventually he calls on one student who has so far been quiet in class, but who correctly answers that the statement means that the p-value in the experiment is less than the chosen threshold of 0.01. This meant that they rejected the null hypothesis of zero impact: the results were highly unlikely to have happened if that had been the case.

Emily, an American student, raises her hand. "I get that if the null hypothesis were true, there'd be less than a 1% chance of getting those results, so we reject the null. But how would we explain it", she asks hesitantly, "to a smart person not well-versed in statistics?"

The class laughs: like Henry earlier, she is using Dan's language. It is a good indicator that key ideas are penetrating the class when students start using the language themselves.

Dan smiles, and allows the laughter to wash over the class. Part of the humour here is that Emily is throwing back a question at Dan, challenging him to come up with a policymaker-friendly explanation on the fly, as he so often does to them. It is an excellent example of students taking responsibility for their own learning in an environment in which they are encouraged to do so.

"Remember the girl with the toothbrush?" he asks. "I'd use something like that. So in this case it would be 'Hey, we set out to assess whether the impact was zero in the population, so we took a sample. The results on that sample are inconsistent with the claim that the impact is zero, so we think there was an impact.'"

Emily's question has helped to focus the class once again on being able to explain statistical concepts without resorting to technical jargon that no-one else will understand. Dan then moves onto interpretation, putting up a question on the screen:

Based on these results, would you recommend the implementation of this program in other similar villages in Vietnam?

"I said 'yes' for this one", answers one student. "It makes sense given that the two contexts are similar that we would want to expand a successful program."

Dan nods but does not validate her answer, and moves on to Amir, an Indonesian student. He has a different answer: "I would say there's not enough information."

Dan smiles. "Not enough information: the favourite answer of this class", he says, referring to the fact that this is always an available option on his Poll Everywhere questions. "Can you tell me more?"

Amir continues: "I'd first want to look at the cost of the program – if it's more than $100, for example, then even though the result was statistically significant, it wouldn't be cost-effective to expand the program."

Dan agrees: "The key thing here is that this is a question about *practical significance.* So even if the impact is statistically significant, that does not necessarily mean you want to implement that policy. Maybe it's not large enough for you to want to do it: practical significance and statistical significance are two different things."

Given the obsession with finding statistical significance in results, it can be tempting to declare success as soon as this happens. The reality of many experiments is that even when statistically significant results *are* found, there are questions over whether they will work at a larger scale. A successful research paper does not mean a successful program.

He pauses, and then continues. "Let me tell you why this is important. Academics are *obsessed* with asterisks*. They'll produce a number with an asterisk, and for many of them, this is the end of the story."

He looks up at the class. "You? No, no, no. You should be asking yourself whether the effect is large enough to warrant anyone doing anything different in the world. If an academic comes to your conference, and says, 'this result is statistically significant', and the next words out of their mouth are *not* 'and by the way, the magnitude of this effect is such that XYZ', you should be raising your hand and asking how practically significant the effect is."

He moves onto a second question:

> *Based on these results, would you recommend the implementation of this program in villages in China?*

Alejandra, from Mexico, says that she would hesitate to expand the program to China because of questions over 'external validity'. This is the statistician's way of saying that your intervention worked here, but might not work somewhere else. For those who want to change the world, this can be a disheartening reality.

Dan is highlighting a tension within the field. For him, there is too much focus on 'internal validity': the likelihood that you drew the right conclusion from this specific experiment. This is why he criticises

---

* Asterisks (*) are often used in output tables to indicate statistical significance.

academics and their asterisks: a successful experiment may help to promote the career of the researcher in question, but do nothing to improve the world if it is not accompanied by practical knowhow and an assessment of *external* validity.

For all the scepticism he encourages the class to bring to the world, though, Dan's over-riding message about answering questions with statistics is an optimistic one. He puts up a chart showing the performance of exit polls against presidential election results. It shows that despite the lack of trust in pollsters today, the exit polls are almost always within 2-3 percentage points of the eventual result.

"Polls on average do a pretty good job", he says. "They typically use about 2,000 people to calculate these results. Think about that: the idea that you want to predict the result of an election with 100,000,000 voters, and you can get within 2-3 percentage points by just asking 2,000 of them? That's *magic*. Really remarkable. A hundred million voters, and you get that close. I get *goosebumps!*"

10

THE SEARCH FOR CAUSALITY

You have probably heard the expression 'correlation is not causation' before. A classic example is to show a graph of infant mortality rates against the number of televisions per person in the population. There is a very strong correlation between high infant mortality and low television ownership. However, sending thousands of televisions to poor countries is unlikely to help: we have correlation without causation. A third factor – low income levels – is the key driver behind both infant mortality and fewer televisions.

For most of us, this fallacy is something we can keep at the back of our minds, knowing that one day we might be presented with misleading data. For statisticians and researchers, the search for causality is the heart of the issue. Imagine we implemented a program that gave cash to farmers, and noticed that agricultural outputs were much greater that summer. However, we also saw that rainfall was unusually high. Unless we have designed a careful experiment, we will have no way of knowing whether the greater output was due to the cash program, the weather, or something else.

This class is the first after the midterm exams, and Dan tells the class that he wants to start with a story. "This is an experience", he says, "that shaped my professional life in a very meaningful way. I spent a year at the World Bank during my PhD program, and I was assigned to work on a project to estimate the effect of textbooks on test scores in Kenya."

The students nod along: they are familiar with this kind of study, having seen this kind of thing in their other classes in the program.

"When I showed up", Dan continues, "I didn't understand the question. What could be more basic than textbooks for learning? Only one kid in six had a textbook. To me, the answer to the question 'what is the impact of textbooks on learning' was obvious. Of course they're going to be important!"

Dan pauses for effect. "The experience shaped my life because when we looked at the data, the average effect of textbooks on test scores in Kenya was as follows."

On the screen appears the word 'ZERO' in big letters. The class smiles, having suspected that this was what Dan was building up to. "This shaped a lot of how I see the world. There are lots of good intentions that don't translate into the impact you want. Today's class is about how you measure impact: how you assess what is a causal effect."

We are starting to move towards the field called 'econometrics': the intersection between economics and statistics. Dan has a problem: lots of the class will have taken econometrics courses before in their undergraduate degrees. To illustrate this, he asks for a show of hands to count the number of people who have taken zero, one, two, or more than two econometrics classes. The group splits roughly into four quarters, making the issue clear. There will always be those who feel Dan is going too slowly for them, and others who feel he is going too fast.

"This is my problem, not yours", says Dan, "but I just wanted you to be aware of it." In doing so he takes ownership of the issue, while anticipating the challenges ahead. This also puts responsibility on the students, as they become more inclined to resolve the issue themselves rather than blaming Dan for the pace. Less experienced students know what they are up against, and may choose to do extra preparation before lectures. Students who have seen this before know that some of the early material may not challenge them enough, and they may seek out extra resources, with Dan's encouragement. This is a common theme in Dan's teaching: anticipating issues helps students to focus.

COMPARING APPLES WITH APPLES

To help make clear the logic of causality, Dan wants to introduce the class to the idea of a 'counterfactual': what your world would have looked like in the absence of a particular event. "This idea is *more* than an airport idea", he says. "It goes beyond statistics, and will be helpful for your life in general."

His first example is relevant to every student in the room, and something labour economists devote entire careers to. "How much more do you earn as a result of going to college?" he asks. "What does that phrase, 'having a college degree *causes* higher wages', really mean?"

The point here is that your data might tell you that those who go to college earn, say, 30% more on average than those who do not. But this is not helpful unless we know what they *would* have earned had they not gone to college. Maybe people with high earning potential are more likely to go to college in the first place, and would have earned 30% more anyway: if this is the case, then going to college has no monetary value. The only way to measure the returns to college is to compare college students with a *similar* group who did not go to college.

This can be very difficult: even if you find some way of finding a group who are similar on backgrounds and test scores, there may be things that you cannot observe that cause the two groups to be fundamentally different. Perhaps those who get good test scores but do not go to college are more entrepreneurial: this is almost impossible to quantify. As we discussed in Part II, the fact that learning is invisible makes the returns to education difficult to measure. We are not in the fantasy world in which learning is immediately rewarded.

"In this class, and many others, you're going to see lots of complex ways that economists have created to do experiments and make causal claims, based on a control group that does not receive an intervention, and a treatment group that does. Whether or not they are valid often comes down to this one thing: *how well the control group mirrors the counterfactual.*"

On the side screens, he has a picture of two groups of bright green apples, and explains that this is akin to 'comparing apples with apples'. Finding causality means comparing a treatment group with a control group. For any comparison to be valid, you need to be comparing apples with apples: if a friend says that your apples are tastier than my oranges, you do not know whether it is because you are the better fruit-grower or because your friend prefers apples to oranges. Similarly, you cannot claim to have found a causal effect if the treatment group and the control group are fundamentally different in the first place.

To drill down on the concept of a counterfactual, Dan asks the class to chat for a couple of minutes to the person next to them about what they would have done if they had not gone to college after high school. "In a few minutes", he says, "I'm going to ask if anyone heard interesting things", getting a nervous laugh from the students.

After some time has passed, he goes round the class. One would have become a swimming instructor; another played squash for India in her age group and might have turned professional. Kosuke, from Japan, says that it is clear from his conversation with Virat that he was an excellent guitar player. "If Virat had not gone to college, he might have entered the music industry and toured around the world."

The class laughs: not many of them knew about this hidden side of Virat before. "So Virat's counterfactual", Dan replies, "is that he would have been a rock star."

"You have very different plans for yourselves in that counterfactual world. I want you to have that counterfactual in mind as you think through this part of the course. When we talked about decision analysis, we said *'don't judge the quality of the decision by the quality of the outcome'*. That often means thinking about the counterfactual:

what would have happened had I *not* done this, and what information did I have at the time?"

In the next class, Dan will put up a heavily-photoshopped picture of Virat as a rock star with an electric guitar up on the screen. It is a bizarre image, but a memorable one. The goal – as with the little girl and the toothbrush, and the groups of green apples – is to give the students an image that the brain can latch onto when the idea of a counterfactual comes up in the future. At various points in the remainder of the course, Virat and his guitar will appear on screen when Dan wants to remind the students of the idea.

Of course, the nature of the counterfactual is that we can never observe it. "If we could, I would be out of a job and you guys would have nothing to do for the rest of the semester", jokes Dan. If Virat really *would* have been a rock star had he not gone to college, then going to college was probably a bad idea. He will never know. And yet, economists and historians must make judgments all the time in the absence of a counterfactual[*].

Dan continues with the example of the returns to education. On the screen is the following question:

---

[*] Tony Blair, the former British prime minister, often points to this in defending the decision to go to war in Iraq in 2003: how could opponents *know* it was a bad decision given that we cannot observe the counterfactual? The reality is somewhere in the middle: no-one can know for sure, but one can still 'be a Bayesian' and update their priors based on new information.

Assume we asked people of similar age how much they earned per hour, and how many years of education they had. We plot the results, and find the following relationship on average:

$wages = 3.5 + 0.2 \times years\ of\ education.$

What does the 0.2 represent?

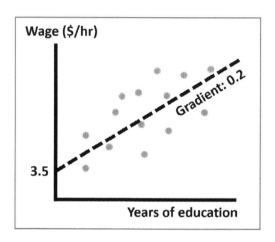

The students do their best to answer precisely what the figure of 0.2 means in the context of the equation. Tsetseg, from Mongolia, goes first: "When you have an additional year of schooling, your wage increases by $0.20 per hour."

Without validating Tsetseg's response either way, Dan looks around the class for more ideas and contributions. Ling, from the U.S., says: "I would just add that the $0.20 increase is *on average*. It's not the same for each person."

Emmanuel, from Côte d'Ivoire, sounds like an economist: "I would say that the *marginal contribution* of one year of schooling, on average, is $0.20 per hour."

Monica, from Mexico, has a point to make about causality. "I think it's important to say that the figure of 0.2 is just the *association*. One more year of education is *associated* with an increase of $0.20 per hour in wages, on average. We don't know that it caused the increase."

Dan smiles, and tells the class: "I am going to be pedantic here. I want us to be extra-careful with language. It's not because it's some weird academic thing, it's because the language we use will dictate what we think is going on."

He pauses. "The first two interpretations above sound like *causal* language. 'When you have an additional year of schooling, *this* happens as a result'. Monica is right to point this out. Every year in this class, we name someone the 'causality police'. This year, it will be Monica."

The class laughs, and this is a prime example of students being given responsibility for their own learning. Monica has been given a role of her own within the classroom, and asked to police the language of the other students.

"Whenever any of you remotely interpret something as causal that is *not* causal, Monica is going to come to the rescue. Our default should be to interpret this kind of equation as a simple association, and only when we have evidence to the contrary should we start talking about causal effects."

As Monica and Dan are highlighting, all we really have is an *association* between schooling and wages. This is the word economists use when they have correlation between two things but no evidence of causation. Similarly, there is an association between low television ownership and high infant mortality, but no suggestion that one causes the other.

Later in the class, Rajiv, who is one of the top students, accidentally describes one of these correlations as an 'effect'. Monica instantly stops him, in the manner of a police officer catching a criminal in the act, and tells him he needs to call it an 'association' instead. This brings the house down: one of the many benefits of giving responsibility to the students is that they love to see it exercised.

THE 'GOLD STANDARD': RANDOMIZED CONTROL TRIALS

At this point, it might be tempting for economists to throw their hands up and admit defeat: if we can never observe the counterfactual, why bother doing research at all when we have no idea whether one thing has a causal effect on another?

Recall that not observing the counterfactual is a problem because it means we cannot compare apples with apples. When we wanted to know the impact on wages of going to college, we could not just look at the average wage of former college students and compare with the average wage of those who did not attend college. People who go to college have different traits, on average, to those who do not. We would be comparing apples with oranges.

There is an elegant solution to the problem of constructing similar groups. If we start with a group of people, and *randomly* assign each one into two subgroups, then those two groups will on average be similar on any given characteristic. If the larger group was half-male and half-female, then when we assign people randomly into the two subgroups, those subgroups will *on average* also be half-male and half-female. It is unlikely to be exactly 50-50 for any given randomization, but there is no bias in play.

The beauty of randomization is that this similarity on average is not just for things you can observe (e.g. age, sex, income). It also applies to things you cannot observe (e.g. level of motivation to study). On average, people will be equally motivated in both groups. Experiments in which a larger group is split into treatment and control group by randomization are known as *randomized control trials*, often abbreviated as 'RCTs'.

RCTs allow researchers to compare apples with apples. This is crucial for finding causal effects: the logic says that if we find a difference between the two groups after the experiment is done, that difference is likely to have arisen *as a result of* the experiment because the two groups should have been roughly similar to start off with. This methodological advantage has led some to call them the 'gold standard' of experimental work.

However, there is debate among economists over whether they deserve this title. Critics say that the field has become so obsessed with *internal* validity (whether they can prove the experiment worked) that they have forgotten about *external* validity (whether it will also work in a different context). Others have pointed to RCTs as unethical, particularly when groups of disadvantaged people are randomly assigned into a treatment group that could make them even worse off. A highly controversial study in 2020 looked at the impact of disconnecting poor families in Kenya's slums from their water supply[xxviii]: the treatment group was threatened with their water being cut off, and the control group was not. With justification, many were outraged about the treatment of human beings as guinea pigs to be experimented on.

In the seventh problem set, students are invited to write their opinions on RCTs, so that Dan has an idea how much of this debate the class has already been exposed to. Some are learning about them for the first time, but others come with opinions ranging from "RCTs are the most significant advance in recent memory in the field of international development" to "RCTs are a total waste of time".

Dan asks the class for a volunteer with experience in the field to explain the basics, and Federico obliges. He uses the analogy of medical testing, in which treatment and control groups are randomly assigned, and then the new drug is given to the treatment group, and a placebo is given to the control group.

Dan nods. "The key thing is that if you see differences in outcomes between the two groups, then that difference is going to be due to the treatment and not due to something else. You have eliminated the possible explanation that the two groups were different to begin with."

Steven has spotted a flaw in Dan's logic. "But the two groups aren't going to be *exactly* the same, right? On any given metric, you'd expect them to be significantly different 5% of the time, even though the groups have been created randomly. How do you know that differences in outcomes are not because you got unlucky with the randomization, and accidentally created two significantly different groups?"

Dan nods, and looks at the class. "This is a really important point", he begins. "Let's pause for a second and take a look at a famous experiment from a few years ago."

Project STAR[xxix] – short for 'student-teacher achievement ratio' – was an RCT carried out in Tennessee in 1996. A few of the class are familiar with the experiment, and Dan asks Rajiv to explain to the rest of the class.

Rajiv begins: "They were trying to test the impact of class size on test scores. They randomly assigned students, by classroom, into a control group with a normal student-to-teacher ratio, and a treatment group with a lower student-to-teacher ratio. They were wondering whether lower class sizes would mean better exam results."

Dan puts up a table related to Steven's question. "The first thing they did", he says, "is to verify whether the randomization worked."

The table shows hypothesis tests of three different variables: poverty, sex, and race. The tests all show there is no statistically significant difference between the treatment and control groups in any of these three indicators[*]. This sounds like a good start, as it suggests they are comparing apples with apples. However, it is possible that the two groups were different on some other variables that they did not test for. Dan moves to a question on Poll Everywhere: "Did the randomization work?"

The students respond using their phones, but Dan does not show the results. Instead, he asks the class to get into small groups and try to

---

[*] In the language of hypothesis testing that the students are now familiar with, for each of the three variables they failed to reject the null hypothesis that the two groups are similar, since the p-values were all bigger than 0.05.

persuade each other of their answers. The question provokes debate within the groups, and Dan lets this debate run for a few minutes before asking them to respond to the same question for a second time.

Looking at the results again, he addresses the class. "I can see that some of you changed your minds, so I'd like to hear from those people. Can someone tell me why their answer changed?"

Ana, from Spain, was one of those people: "We had no information at all about grades, so I originally put 'not enough information'. But thinking about it again, you're always going to have that problem – you can't guarantee the two groups will be exactly the same for everything. It looks like the randomization worked on the variables they were interested in."

Dan nods, and adds: "It sounds like you wanted information on the test scores each student was getting at the start of the experiment, just to make sure the 'randomization' didn't produce one group that was already smarter than the other. It's a good point: why is that information not in the table?"

A couple of students suggest that the information might not have been available; a third says that perhaps it was available but the researchers believed poverty, sex and race to be the most important factors.

Dan agrees with these possible explanations, but adds: "for those of you who have been involved in RCTs, perhaps they didn't have *time* to find out the test scores at the start of the experiment." Once again, Dan is moving the class from theoretical explanations to practical ones. Researchers are always short on time and money.

He asks another question. "Can anyone think of a more *malicious* explanation as to why they might not have included baseline test scores in this table?"

This gets the students interested. Emily suggests that they might want to hide a flaw in their experiment. Chris then points out: "Maybe they're cherry-picking, and they only showed the tests that said that randomization worked."

Dan has been waiting for this: his next slide is the colourful picture of the girl picking cherries from a few classes ago. Project STAR is a well-known study, and it is not Dan's intention to accuse its authors of cherry-picking. But he wants his students to return to the world with a sceptical eye: when something expected does not appear, he wants them to ask, 'why not?'*.

In the debate about the effectiveness of RCTs, Dan describes his views as being 'closer to positive than negative' on the spectrum. Given that this debate can turn political, he is open with the students about his beliefs. But he is also determined to ensure that the students do not see RCTs as the 'gold standard' they are sometimes described as. It is difficult to compare apples with apples – that is, for the control group to mimic the counterfactual – even when you create the control group by randomization.

If the randomization *does* work, and the RCT has been properly carried out, then researchers can validly claim to have found causality

---

* This is the academic's version of the "curious incident of the dog in the night-time", from an Arthur Conan Doyle story in which Sherlock Holmes deduces that the *lack* of noise from the dog suggested the unknown midnight visitor was not a stranger to it.

in their results. They will do the apples-to-apples comparison between the treatment and control group, and hope that the treatment received leads to statistically significant results in the outcome they are interested in. In the case of Project STAR, they found that smaller class sizes *did* lead to significantly higher test scores, and the paper has been influential in education circles ever since. There is theory involved, but as usual in statistics, the search for causality can often be as much an art as a science.

THE ECONOMIST'S CURSE: OMITTED VARIABLE BIAS

During the classes on causality, the students learn about a tool called *regression*. When you have a scatter-plot of points on a graph (like the one in the previous section with education and wages), a regression is used to find the 'line of best fit' that you can trace through the points. In our example, the equation of the line was *wages* = 3.5 + 0.2 × *years of education*, which indicated that the slope of the line was 0.2, and the intercept with the y-axis was 3.5.

The students learn how to extend the tool into using multiple variables. For example, if there is a gender wage gap, a regression including a variable that equals 1 for males and 0 for females might give you an equation looking like

wages = 3.0 + 0.2 × years of education + 1.0 × male.

This would indicate that a man with 10 years of education would expect to earn $6 per hour, while a woman with the same education would only expect to earn $5 per hour. Regressions are what statisticians use to find the numbers in equations like the one above.

In addition to finding these numbers, a regression will also return a figure called the R-squared, which indicates how well the line fits the data. If the points are all close to the line, the R-squared will be close to 1; if the points are all over the place with no correlation then it will be close to 0.

You may already have a good sense for why correlation does not mean causation: sending televisions to poor countries will not increase the infant survival rate, because both high television ownership and high infant survival are caused by a third variable of high income. Economists call this 'omitted variable bias': in seeking to explain variation in infant survival rates, we are falsely attributing to television ownership what is really attributable to income. The 'effect' of television ownership on infant survival will be biased upwards because we have omitted income from our regression.

Dan's example also uses income as the omitted variable, and looks at the Project STAR experiment again. He asks the students to imagine that the authors of the study had done the following regression:

test scores = 100 – 1.5 × student/teacher ratio

In this example, the more students per teacher in a school, the lower the average test score would be expected to be. However, we have an omitted variable problem. Kids in high-income families are more likely to go to schools with fewer students per class. However, they are also more likely to have lots of other resources available to them for learning, and their parents are more likely to be well-educated themselves. How do we distinguish between class size and income when seeking to explain variation in test scores?

Dan shows them the regression *with* higher household income included:

test scores = 100 – 0.3 × student/teacher ratio – 0.7 × income

He asks someone to interpret the way the coefficient on the 'student-teacher ratio' variable changed between the two regressions. John begins: "Since the 0.3 is less than the 1.5 in the first regression, we're over-estimating the effect of the..."

"No, no, no, no!", interrupts Monica, with comic vigilantism, from the other side of the room. Our causal police has been listening in, ready to act. Startled, Dan turns round, and sees who has spoken. A smile comes to his face as he realises what has happened: he nods at her as the class laughs, and clenches his fist victoriously. He has been ready for this moment, and quickly puts a picture of Monica on the screen, wearing a police officer's hat, that she had taken at a recent party to celebrate her new role. This, in turn, gets a lot of laughs from the class.

John had said 'effect' instead of 'association', asserting that there was a causal link between small class sizes and higher test scores. We already saw that without controlling for income, any claim of causality is flawed. Even when you *do* control for other variables, you can never be sure that you have not missed anything. Perhaps there is some greater factor, like the strength of political institutions (e.g. the rule of law, and property rights), that in turn drives income and test scores. Omitted variable bias is impossible to avoid without randomization. However, if you do not care about causality, and only want to make predictions based on the data you have, there is still plenty of analysis you can do.

11

PREDICTION AND MACHINE LEARNING

Of all the topics the students will cover in API-209, machine learning is the one that sets the most pulses racing. They know that these few classes could be a gateway to some of the most exciting things happening in technology across the world: big data, artificial intelligence, self-driving cars, smart cities, robotics. With the increase in processing power in recent decades, plenty of new statistical techniques are possible. Even for those that will not become data scientists, there will be an increasing need in many places of work to understand the key concepts behind machine learning. Dan hopes to channel the enthusiasm for the expanding field into his class.

"Today is a very exciting class", he begins. "The topic is exciting, many of you have worked in it, and the future of a lot of work will be done in this area. The class is about prediction. For some of you it will be very basic, but for many it will open a new world. I predict that some years from now, a Nobel Prize for economics will reward this work."

He has the students' attention. "You might remember, when you were all young and energetic, that we talked about three uses of statistical inference: *descriptive, causal,* and *predictive.* We've spent the last few classes on causality: what is the effect of X on Y? Prediction is simply 'what is the predicted value of Y given X?'.

The key difference, Dan explains, is that they are no longer looking for a causal link. "I want to start with an example", he says. On the screen is a picture of a box of some small, circular pieces of fabric. "Does anyone know what this is?"

Amy, an American student, puts her hand up. "Those are things you get from IKEA to put underneath your chair to stop it from scratching your floor."

"Do you own these, Amy?" says Dan, smiling.

"*Maybe*", she replies, to laughs from the class.

"It turns out", Dan continues, "that people who buy these things tend to be very good at paying off their credit cards. If you're a credit card company, and you want to know if you'll be paid back, you might be interested in whether they bought one of these. Let me ask you a question. Do you think there's a causal link between buying this thing, and paying off your credit card?"

"No", says the class in unison. People do not suddenly become more reliable bill-payers as a *result* of buying the pads, but it might be a signal of higher risk aversion.

"Even so, for the credit card company, that causal link is not important. They just need to know that people who buy these pads are more likely to pay them back. So if you want a couple of tips: first, buy those scratch things. Second, don't check your statement at two in the morning. By the way, prediction is also at the heart of fraud prevention: they see an unusual transaction, and it has an associated probability of fraud."

Dan pauses. "I'll say one thing at the outset: economists have been *obsessed* with causality. You go to a seminar and it's all they're talking about. But there are many, many important questions in the world where all you need is a prediction: you don't need a causal effect."

He goes back to the example from the week before about the returns to education. The students were taught that if they see an equation like *wages* = 3.5 + 0.2 × *years of education*, they cannot interpret the 0.2 as a causal effect of education on wages. However, they *can* use it to predict, and a student correctly answers a question asking them to predict the hourly wage of a person with ten years of education[*].

Having brought up education, he asks Juliana, the Brazilian student who spoke about her teacher training program in the first lecture, to tell the class about her experiences with prediction. "I was working for a company that runs schools in Brazil, and we had a high dropout rate. We used machine learning to rank students from highest to lowest probability of dropping out, to help education co-ordinators to focus their efforts."

---

[*] This is $5.50: substituting '10' into 'years of education', we get *wages* = 3.5 + 0.2 × 10 = 5.5.

Dan asks: "Did you get a sense of what factors were driving it?"

"It's difficult to know exactly", says Juliana, "but there was a big peer effect: if friends dropped out, it made you more likely to drop out. If parents were involved, and checked their results online, that helped. But we saw a pattern where they would enter the portal in the first few weeks and then stopped suddenly."

"This is a great example", nods Dan. "Look at the use of data: they knew whenever the parents were accessing the portal, and all that went into the database. Some of these things might be causal, others not – but either way it allowed her company to make better decisions in spending its scarce resources."

A second student, Shreya, also used machine learning to predict student drop-out in India. A third, Charlotte, used to work for a telecoms company, and used machine learning to predict the factors that led to customers switching out of their service: younger people switched more, as did those who owned certain products. They then focused their efforts on the group identified as at a high risk of switching, as it is cheaper to retain existing customers than attract new ones.

"No wonder I haven't got a single phone call", laughs Dan.

Despite plenty more student experiences he could go to, his notes from last year's class said that he spent too long motivating and not enough time in the main body of the class. He tells the students this, bringing them into his thought process.

"Machine learning means using computer algorithms to make predictions", he continues. "The machine learns from the data collected to improve the algorithm. Let's say you're interested in predicting house prices, and you have data on 50,000 houses. What do you think are the predictors for the price of a house?"

The students fire out answers. "The size of the house", says one.

"The last price at which it was sold."

"How old the house is."

"The location of the house."

"The price of houses nearby."

"Whether there are good parks and schools around."

Dan collects these ideas on his iPad, which is linked to the screen. "We can come up with a very long list: some of these we'll have data for, others we won't. We'll use it to predict. Any ideas, based on what you already know, on what you could do with the data you have?"

One student suggests that they could use a regression to help them predict house prices given the data that the class just came up with. Dan nods. "That's great - you already have a tool that will help you create a model. What are the drawbacks to this method?"

Ana, from Spain, points out that regressions do not tend to work well when there are too many variables. When you have so much data, it becomes harder to be sure what the movements in house prices are attributable to. This is even more true when your variables are correlated with each other, as many of the list identified by the class would be[*]. When two variables 'look like' each other, it's hard to say which one is the driving force behind house price movements.

Salma, from Morocco, has spotted another drawback. "With your example of the scratch pads, that's not something you would *think* of to add to the regression if you were trying to predict creditworthiness. Part of the value in machine learning should be that it will tell *us* what's important, and if we rely on ourselves to come up with the list of variables, we won't think of the surprising ones."

Dan jumps in. "This is a good point. How do we know if our predictions are any good? We do have a measure of how well the model fits the data. Why don't we look at the R-squared?"

Salma, who has some experience in machine learning, replies: "You might find that you get a high R-squared, so you've done well at predicting prices for these 50,000 houses. But you already know the prices of these houses because they're part of the dataset. You don't know how good your model is on *new* data, which is kind of the idea."

Rajiv chimes in: "There's likely to be some noise in the data, and if you train your model to find the line that's closest to all the points, you will accidentally pick up the noise as if it were the real underlying signal."

Dan nods. "The problem that Salma and Rajiv have identified", he says, "is called *overfitting*." He puts up a graph on the screen, shown overleaf, to illustrate it.

"Which of these two models fits the points better?" he asks.

---

[*] For example, being in a 'good' location is likely to be correlated with the proximity of parks and schools, and the last price sold will be correlated with the prices of other houses in the area.

"The curved one", the class says in unison.

"Should we *use* the curved one?" he asks.

"No!" they say.

"Why not?" he asks.

Beza, from Ethiopia, puts her hand up. "The reason we do prediction is that we want to apply our techniques to data we haven't seen, so if we overfit to data we're seeing right now, it probably won't work for the new data."

Figure 11.1: we have a choice of models to fit the points in the graph. The curved solid line fits the points almost perfectly, but will probably perform worse than the straight dotted line when we try to apply the model to new data. The dotted line may capture the underlying signal, whereas the solid line also captures the noise in the data. This is an example of 'overfitting'.

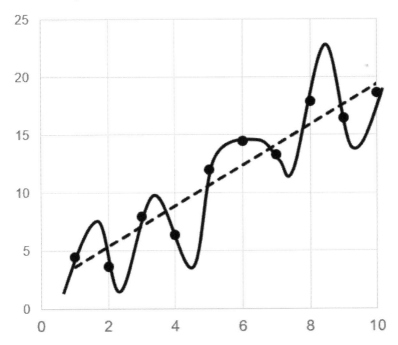

"So it looks like we have a problem", says Dan. "How do you build a model that is flexible enough to capture the nuances in the data, but not *so* flexible that it overfits the data?"

Salma knows the answer: "We can split the dataset."

Dan continues: "Great idea. We will divide the dataset into two. We'll take 70% of the data at random, and call this the 'training set'; the remaining 30% will be the 'holdout set', which we won't get to see until right at the end."

This is one of the central tenets of machine learning. Just like a professional soccer player, who works on their game in training but is judged by performances on the field of play, machine learning algorithms go through a two-stage process. Having split the dataset in two, you use the training set to come up with various models and then tweak them, trying to get them close to the data. Right at the end, you test the models on the holdout set, and you look at how well it predicted outcomes. The key is to measure the performance of the model on the holdout set, not the training set: a striker is judged by how many goals she scores in matches, not in training.

CASE STUDY: REDUCING INJURIES IN THE WORKPLACE

There are only two classes on prediction: as Dan highlighted, the goal is not to turn the students into experts, but to give them enough understanding for them to be able to ask the right questions in the real world, or to bridge to more advanced classes if they are interested. Between the two classes, the students have a difficult problem set to do, in which they use machine learning to help a government agency to reduce the risk of injuries in the workplace.

At the start of the second class on prediction, Dan recognises the work the students have done. "I've always had the impression that you guys work really hard", he begins. "That impression was solidified when I woke up at five a.m., as I always do, and I took a look at the times at which you submitted the last problem set."

He puts a chart up on the screen showing the times of submission. This is another example of the data that Dan collects in class turning an invisible concept – hard work – visible. It gives weight to the recognition of the students: Dan is not just throwing out platitudes. This allows him to set tougher problem sets, as the students are more likely to manage the workload if they feel recognised. As always, his goal is to maximise learning, not satisfaction, but this is an example of the two goals moving in the same direction.

Figure 11.2: At the start of Class 21, Dan surprises the students with a chart of their submission times for the machine learning problem set.

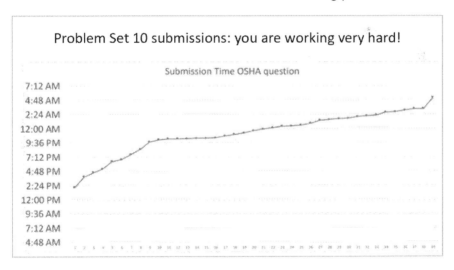

In the problem set, they are hired by the Occupational Safety and Health Administration (OSHA), a U.S. government agency whose goal is to minimise injuries in the workplace. The agency has limited budget and can only inspect 30% of workplaces. Its current strategy is to randomise the list of workplaces that it inspects, but it wants the students to help them to better target workplaces that might be at higher risk of injuries.

The students use three different methods to do this. The first is to run a simple regression, as they are used to doing with other tasks. The second is to 'throw the kitchen sink' at the problem by running a regression that includes every single variable they have, despite warnings of overfitting. The third is a machine learning algorithm that tries different variables but assigns a 'cost' to each one, meaning that it will naturally prefer to use fewer variables to avoid overfitting.

They find that the 'kitchen sink' method, as they expected, overfits the data: it performs well in the training set but poorly when it is applied to new data in the holdout set. The machine learning algorithm is the best of the three in predicting high-risk workplace sites, but it is not a huge amount better than the simple regression. All three are much better than the current strategy of picking sites at random.

Emily takes up the challenge of explaining this to a policymaker: "Before, out of every 100 sites we inspected, only 44 were actually dangerous. With the machine learning algorithm, for every 100 sites we inspected, 78 of them were dangerous. We've got much better at predicting high-risk sites."

There is still some debate among the class over which they should recommend to OSHA. As always, there is more to it than picking the most 'technically correct' method. Machine learning can be something of a 'black box', and the simple regression method was good enough to mean that 75% of inspected sites would be dangerous. Given that this is only just below the machine learning success rate of 78%, and well above the status quo figure of 44%, we might recommend to OSHA the simple regression method, whose inputs and outputs can be understood by everyone.

Summing up, Dan advises the class: "Don't waste time searching for a perfect prediction: a better question to ask is 'can I do better than the status quo?'. In the real world, often a small improvement in the status quo is the best you can hope for."

By the end of the two classes on prediction, the group is approaching the end of the semester. There are only a couple of weeks of classes left, and they have covered the key ideas of probability, hypothesis testing, causality and prediction. They know how to answer questions with statistics, and they are good at explaining their conclusions to policymakers, like they will have to do in the real world. We will end the book with the final lectures of the class, but now we return to the invisible dynamics that Dan navigates in his teaching.

PART IV: THE INVISIBILITY OF TEACHING

12

THE BATTERY LEVEL OF AUTHORITY

We said in Part II that the learning environment is the collection of invisible bonds between the members of the class. We distinguished between horizontal bonds, between students, and vertical bonds, between students and teacher. We also described some of the ways that Dan has strengthened these bonds. However, we have not yet defined what these 'bonds' are, other than to say they are invisible.

At a fundamental level, we have a bond with someone else when we trust that they will provide a service that we are looking for. We are all looking for a multitude of services from other people, some of which relate to our innermost hungers and desires. As well as expecting Dan to provide the service of teaching us statistics, we also want to feel safe in his class. We may also want to feel important, or powerful, or seen, or loved.

The notion of trust as 'service provision' seems transactional, but we do pick our friends based on how they make us feel. By watching soccer with a sports-mad friend, I am providing her with a service. Another friend might want to go to a nightclub every Saturday night. In admitting to him that I dislike loud music and would prefer to stay at home, I may be failing to provide a service that he is looking for. In time, I will gravitate towards the friends who will watch soccer with me, and away from the ones who I worry will drag me out to a nightclub. Like everyone, I seek out those who provide the services I want.

Heifetz has a name for this trust: *authority*. We 'authorise' people to provide services, and the more we authorise them, the stronger the bond[xxx]. There are two types of authority: 'formal' and 'informal'. Formal authority is bestowed by a higher power: Dan is in a position of formal authority over the students because the university is employing him as a professor. This is a fixed concept: the university can promote him, or decide not to employ him anymore, but until then his formal authority does not change.

A person's informal authority changes all the time depending on whether they provide services we are looking for. Imagine someone who puts their hand up in class to make an interesting point. This is a helpful service to provide, and their informal authority may rise as a result. But having made the point, they go on and on, speaking for over a minute. If we expected them to provide the service of 'being concise', then their informal authority may fall[*]. Formal and informal authority

---

[*] In theory, informal authority should be task-specific: if we know that a person is smart but inconcise, we should trust him to make good points but not to wrap them up quickly. In practice, much of this happens subconsciously, and a loss of informal authority in one task can lead us not to trust a person in other tasks.

are not always correlated: a bad professor may have high formal authority but low informal authority with the class, whereas it might be vice versa for a smart student.

Bonds strengthen when the members of the classroom *authorise* each other. To trust that someone will provide services you are looking for, it helps that they have reliably provided those services before. Each time Dan makes his students laugh, or feel confident, or learn something, he improves his track record in providing important services and strengthens the vertical bonds of trust. The same is true for the horizontal bonds between the students.

Just as the bonds are invisible, so is the level of authority that we have with each other. Dr. Kim Leary, who teaches a class on authority at the Kennedy School based on Heifetz's framework, came up with the metaphor of informal authority as a 'battery level'. Each person can imagine a battery level indicator appearing above everyone else in the class, which starts to drain whenever they start talking. The higher someone's authority, the more battery life they have, and the slower it takes to drain. You can probably picture the person in your class who had low authority: whenever they started to speak, you rolled your eyes and switched off. Their battery level started at 'red' and drained almost immediately. In contrast, Dan's authority, helped by his formal position as the professor, starts at 'green', and students authorise him to talk more than others.

In some ways, life would be much easier if we could see our own battery levels. Those who struggle from 'impostor syndrome' imagine their own battery levels as lower than they are in real life, and may be shy to speak up for fear of the battery draining. We may have the same issue when meeting a celebrity we admire, or the CEO of the company:

we imagine they will not want to listen to us, and perceive a rapidly-draining battery, causing us to fluff our words. In contrast, those who are over-entitled perceive higher battery levels than they have in real life. They keep on speaking, assuming their audience is rapt with attention, when in reality they have tuned out as the battery expired.

These 'battery levels' are real but invisible: Dan cannot see his own, or the extent to which it varies by student. Even if most students respect him as a teacher, they still have a finite attention span. He will have to use his experience, take the temperature in the room, and react accordingly if he feels the battery is running low. In getting the students to take responsibility for their own learning, he aims to strengthen the horizontal bonds by getting them to authorize each other. This often means increasing the battery levels of the students, so that they feel entitled to intervene in class and teach each other. As we discussed in Chapter 7, this can only be done in an environment of high psychological safety: if students fear failure, they are likely to perceive their own battery levels to be weak. They will contribute less as a result, and lose an opportunity to learn. The rest of the class, too, will lose an opportunity to learn from them.

AUTHORITY AS A RESOURCE: DR. PARSONS AND STEVE BUCHANAN

In *Leadership Without Easy Answers,* Heifetz describes the way that a position of formal authority, like the one Dan has in the classroom, is both a resource and a constraint. By virtue of being the professor, Dan can do things that no-one else in the classroom can. However, great power comes with great responsibility, and the expectations that weigh on figures of authority can make it hard for them to focus on their purpose.

Heifetz tells the story of Dr. Barbara Parsons, an American physician[xxxi]. In 1985, her 42-year-old patient, Steve Buchanan, called her complaining of another kidney stone. Sadly, scans showed that he had stomach cancer instead. Steve was a carpenter, physically strong, and the proud father of three children. The potential for a slow deterioration into weakness – and possibly even his own death – was the ultimate adaptive challenge to confront. Dr. Parsons led them through the whole process, giving them information when she thought they were ready to hear it, and helping the family to navigate difficult issues around illness and death.

Parsons was in a position of authority, and used it to help Steve to manage his new situation. Between Steve, his wife, his children, and herself, she created a holding environment in which her personal warmth and expertise made them feel like she had their arms around them. She could direct attention: the family would look to her for a diagnosis, and follow any plan she set out. She had access to information: not just the results of the scans, but also the clues in the words and body language of Steve and his wife. She could use those clues to sequence the flow of information to the family, based on her assessment of their resilience.

For example, Steve's reaction after having surgery to investigate the cancer was to say "This can't be real. I'm not ready for this." His wife asked, "Tell us the good news, Doctor: Steve's going to be okay, isn't he?" The results of the surgery were bad: the cancer had spread, and privately, Dr. Parsons was heartbroken. But she also needed to contain the anxiety of the family to minimise the risk of a downward spiral into depression, so she sequenced the flow of information, telling the truth but not the whole truth, to give them time to develop the ability to respond. She looked for small clues in their language that would hint they were ready to confront the next stage of the challenge.

She also had the power to frame issues around death: to speak about them in a way that influenced the discussions the family would have among themselves. She could orchestrate conflict and disorder if she needed to, subtly encouraging the family to have difficult conversations she believed were necessary. Finally, as the authority figure, she had some power over choosing the decision-making process: depending on what she thought was best, she could take some medical decisions in Steve's best interests autocratically, or insist on getting the buy-in of the family first.

Fourteen months after Steve's initial phone call to Dr. Parsons, he died at home, surrounded by his family. The story was tragic, but there was also a sense of achievement for Dr. Parsons: she had helped the family to ensure everything was in order, and made sure Steve's wife and children were able to spend as much time with him as possible in his final year. As Heifetz writes[xxxii]:

> *"Parsons helped the Buchanans take responsibility. She neither shielded them from their problems, nor did she abandon them... A large measure of the Buchanans' work involved emotional learning... Parsons began with the assumption that the family had the potential – the basic capacity – to take responsibility in this new situation, but likely would need help to employ it."*

The same is true, in different circumstances, in Dan's class. Both involve an authority figure working through an adaptive challenge with a group. Heifetz highlights that Dr. Parsons' story suggests seven resources that authority figures have in confronting adaptive challenges: creating a holding environment, directing attention, accessing information, sequencing the flow of information, framing issues, orchestrating conflict, and choosing the decision-making process. Dan is also in a role of helping others to take responsibility, and those seven resources also apply to him.

Creating a holding environment

We defined the 'learning environment' earlier as the holding environment of the classroom, built up of all the bonds between the members of the class. We argued that the strongest learning environment was one in which students acted as a team, with the same common purpose to maximise learning. This is boosted if psychological safety, the key determinant of high team performance, is strong.

The concept of 'team learning' is one in which the teacher gives the responsibility back to the students for their own learning. However, this is not an anarchy. There is fascinating work on the study of leadership in teams like Orpheus, the chamber orchestra without a conductor which nevertheless plays beautiful music. This is not like that: Dan has a crucial role to play as the authority figure at the front of the class. It is his job to create a learning environment in which students feel empowered to take responsibility. The teacher has a great deal of power in the classroom in determining whether invisible bonds are weak or strong.

Directing attention

As the captain of the sailboat, Dan needs to direct it towards the island, despite the winds that are acting upon it. We saw before how there were two ways he could do this: by using the sails he can take advantage of the wind, and by turning on the motor he can push the boat on through despite them. Attention, just like the fuel of the boat, is a currency he uses selectively.

We see this with his 'airport ideas'. These are concepts so important that he wants the class to remember them, five years from now, when they meet him by chance in an airport. To this end, he spends the most amount of currency on them. The phrase "this is an airport idea" leads students to switch back on: they know that this is something they need to listen to. In an ideal world, students would be absorbing every word. No teacher lives in this world, so attention becomes the most valuable currency a teacher has. Dan tries to spend it wisely.

Accessing information

Dan's position allows him to collect data on the class. We have already seen a couple of technological examples: Teachly gives him real data on participation as soon as the class has ended, and insights into any students or groups that are missing out, and Poll Everywhere helps him to collect important data on how well the students understand a concept. This is the information age, and newly available data is pushing the boundaries of what is possible in every profession. Teaching is no exception.

Uniquely, Dan also has access to all the students' problem sets and exams. His course assistants update him each week on which questions the students seem to be struggling with, so that systemic issues can be addressed in class. A goal for the future is to update the course software so that it automatically flags which questions these are, and for individual students, highlights mistakes that are repeated over a period of weeks. The dream is to have the software tell a student "that looks similar to a mistake you made a couple of weeks ago", and propose improvements. Given the fact that student answers are mostly in prose, this will be challenging. It does, though, highlight Dan's enthusiasm for collecting as much data as possible.

A further example of using course assistants to generate data is in time management. Dan splits each of his classes into 6-7 sections, and based on previous experience, he allots a certain amount of time to each one in an Excel file. The assistant is asked to log exactly how much time he spent on each section, so that he can evaluate himself afterwards and decide how much time to budget for the same class next year.

In addition to these sources of 'real' data that can be analysed, Dan is constantly gauging the temperature of the room, trying to read the student reactions. It is usually possible to distinguish between Merton's positive and negative silences from Chapter 6, and work out whether learning is being done, by observation and listening. That said, looks can be deceiving: in motivating the use of Poll Everywhere, he tells a story about early in his career when he ran a poll just to check the class generally understood the idea, and it came out with 17% getting the answer right. Given the time costs of running polls, the key is in striking the right balance.

Dan's data collection is one of the most tangible things that sets him apart from an average professor. He wants to know everything he possibly can about what's happening in his classroom. The more he knows, the more he can react to it, and adjust in real-time. In Chapter 14, we will look in more detail at some of the information he collects through the year, and the way in which he evaluates every class to help him to improve next time.

Sequencing the flow of information

The key use for the data Dan gathers is to enable him to sequence the flow of information appropriately. If it becomes apparent, perhaps through a poll, that the class is struggling with a topic, then he will slow the pace down. If the poll shows that everyone 'gets it', he will speed up. This constant rebalancing helps to ensure he can direct attention where it is most needed. Dan can open and close the flow of information like a valve in order to maintain the optimal level of heat in the room for learning. In general, the more rapid the information flow, the greater the heat. However, going too quickly can cause the whole system to shut down.

Dan also has decisions to make every year about the ordering of material in the class. We saw in the first few classes that he wants to ensure people have a good understanding of basic probability theory before moving onto tougher topics: this is an easy choice. More interesting is his decision to put so much behavioural psychology at the *start* of the class: this is a field that in some branches of economics is seen as an afterthought. Many professors would make sure the material is embedded first, then at the end of the course explore ways in which biases may influence the decisions we make. Dan's course, as we saw earlier, is as much about the relationship *with* statistics as it is about statistics itself. The art of interpreting results is just as important as the science of collecting them.

He sequences the flow of the course to reflect this. In interviews, Dan confirms that in his years of teaching the course he has experimented with many different versions of the sequencing. His biggest challenge, not surprisingly, has been where to put the material on sampling distributions which always causes such a struggle in Class 5. Although technical, experience has shown that it needs to be close to the start of the class: this is one topic that students do not tend to pick up by 'learning by doing', and he believes it is important to understand the concept before moving onto hypothesis testing. "In general it's a case of backward mapping", he says. "What do I want them to be able to do by the end of the class?"

His approach is at odds with some other statistics courses around the world, in which sampling distributions are something like the elephant in the room: students learn how to do hypothesis tests without ever appreciating that they are considering a point on the sampling distribution, and what that means. Dan believes that this is a mistake, and at the centre of his struggles as a PhD student. He chooses to use up a lot of the boat's fuel in understanding this one class, aware that it is an important foundation for what follows. However, the concept is a tough one, and this was borne out when the students struggled to understand it.

Framing issues

Dan has the power to determine how the students view the whole field of statistics. One of our central concerns in the introduction was that statistics is too often framed as a branch of mathematics. Many people have great situational intelligence and common sense, but never thought of themselves as math experts. Dan is clear that much of statistics is about interpretation and judgment: subjective reasoning, not objective truths. This underlies the very first discussion of class, when he gets the students to consider what 29% 'means'. Even a number, the most objective of all things, becomes subjective when our biases come into play.

This influence extends far outside the lecture theatre. Dan's problem sets ask mostly for subjective prose, which generates more discussion outside of class. Often, as we saw with the Mexico case study, he will ask questions to which there is no right or wrong answer. Students will be marked on how well they engaged with the question, rather than whether they got it right. The thinking process is one level up: Dan wants students to talk about how they *thought* about a question, not how they answered it. Similarly, in the final exam, less than half the points are awarded for getting the 'right' answer. Far more are awarded for making the right connections.

Orchestrating conflict

We saw already how Dan turns down the role of 'validator' of student responses. An important consequence of this is that multiple ideas surface in the same discussion, and often do so organically without need for Dan's intervention. A common tactic is for Dan to use the results of a poll to generate conflict in class, asking people to persuade each other of their answers. This conflict is at the essence of team learning: the students will not have all the answers. They will have to reckon with multiple competing views of a problem, and often confront it rather than solve it. This strengthens the horizontal student-to-student bonds in the learning environment, and pushes the students out of their comfort zone.

Note that in this context, conflict does not mean acrimony. Dan wants to expose the students to conflicting approaches and ideas because they often broaden our understanding of a concept. He cites the famous parable of the blind men and the elephant, in which six blind men all touch a different part of an elephant and confidently declare it to be different things. Statistical concepts are the same, and even those who think they understand have something to gain from hearing about the experiences of others.

There is a virtuous cycle between conflict and psychological safety. Student conflict *creates* psychological safety, as students may feel more comfortable being wrong with each other than with the professor in front of the whole class. In turn, psychological safety improves the quality of student debate, as they will put more on the line when they do not fear being humiliated by their peers. If they believe a wrong answer will make them look bad, they are unlikely to accept the challenge of explaining their answer to someone else. The stronger the learning environment, the greater the level of conflict it can withstand.

A great example of conflict engineered by Dan was in the Mexico case study. He knows that every year there will be two factions: the students from the private sector, who think that underspending is a good thing, and those from the public sector, who think it is a bad thing. When Gabriela pointed out that the underspending might be a political issue, Dan fanned the flames by asking why that was such a problem. Gabriela's incredulous response – "not in my country!" had the whole class smiling: some in surprise, others in recognition. By the end of the discussion, the two factions had been brought closer together: those from the private sector had learned something important about the workings of government, and those from the public sector had started to question something they had taken for granted.

For a teacher, orchestrating conflict is not without risks. The expectations of the students will vary: while some might see a debate as helpful, others may be frustrated by the lack of resolution. A common complaint on student evaluations is that class discussion is not resolved to their satisfaction. One student highlights:

> *"There should be more time closing subjects, and more comments from Dan, during class debate. Sometimes when classmates explain opposing views on concepts, Dan doesn't sufficiently clarify what the correct answer was, and why the alternative was mistaken."*

A second student agrees:

> *"I think there was too much class discussion. The teacher would ask us to discuss with our classmates 2-3 times per class, which I think is too much. I think it would be better if he allocated more time to explain the concepts and do examples."*

And a third:

> *"While the focus on creating an engaging classroom environment is commendable, the outcome/takeaway from class discussions was often unclear. There were many times where I left having*

'enjoyed' the class, but without a clear understanding of the concept."

And a fourth:

"Class discussions are fantastic, but it's important to moderate them and bring them back together to clarify concepts. It's easy to get lost in the wide array of things that different individuals in class might have to say!"

There are three key takeaways from these comments. The first is that we need to put them in context: it is sometimes necessary to disappoint expectations of some students. Given that students will have different expectations about how much time is allocated to class discussion, it is inevitable that there will be some who believe there was too much or too little. We have cherry-picked some constructive feedback above, but the background is that when students were asked to rate Dan's management of class discussion on a scale, 64 of 78 rated him at 5/5, and 13 of the remaining 14 rated him at 4/5. This is generally seen as one of Dan's strengths.

The second takeaway, however, is a learning point for Dan: as someone who understands the concepts, it may be clear to him (and the stronger students) in which direction the debate is going. At the end, he may therefore only offer a brief wrap-up, believing that this is all that is required. The first comment above is a helpful reminder that students who are new to the ideas may need a longer wrap-up from him to maximise their learning. Given that these are the students that can learn the most, this is helpful feedback with which he can continue to improve.

The third takeaway is that the failure to provide clarity at the end of a discussion may be linked to the temptation to maximise enjoyment over learning. Students often enjoy class discussions because they are at the centre: in a safe space, they *like* to have responsibility. Dan feels this too, and enjoys the buzz of the classroom when students are supporting each other. But at some point, he also needs to step in and provide clarity if he is to maximise the learning of the students who are hearing about the concept for the first time. Even if most students think he manages class discussions expertly, it is vitally important to listen to those students who disagree: therein lie opportunities for improvement.

Choosing the decision-making process

Just as Dr. Parsons was able to control whether she took decisions autocratically, or with the consultation and consent of the family, Dan can dictate the extent to which the students control the progress of the class. Giving the work back to the students is what team learning is all about, but he is responsible for directing attention and keeping time, getting the boat to move in the right direction at the right speed. Just as a boat's captain might hand over control to the first mate while waters are calm to aid their learning, and then reassume control when a storm appears on the horizon, Dan can give control to his students for an interesting discussion and then take it back when necessary.

AUTHORITY AS A CONSTRAINT

For all the resources that authority provides, there is a single constraint which can often paralyze those in positions of authority from exercising leadership. Authority comes with expectations, and these differ by stakeholder. Here, the stakeholders are not just the students, who form different factions on different issues, but also those in charge of the faculty at the Kennedy School and the wider Harvard machine.

In Dan's case, there is a central expectation that he will teach statistics to the students, but every student will have a different expectation of how quickly he should move through material. Stronger students will want him to move faster; weaker students will want him to move slower. In picking a pace that balances these expectations, he is forced to disappoint most of the class.

Personality and cultural backgrounds may determine students' expectations of how Dan teaches. Some may expect an entertainer; others may see the use of humour as irrelevant showmanship. Likewise, the students will have different expectations about *what* they are going to learn. Despite his best efforts, those who want to learn about policymaking may be disappointed during the mathematical interludes, or frustrated at having to learn to code as part of the class. Those who want more rigour may be frustrated by the lack of mathematical proofs. Those who think behavioural science deserved its Nobel Prize in 2017 may be delighted by his approach to the early part of the course; those who think it a fad will be disappointed. Dan has no chance of pleasing everyone.

The students can threaten Dan's authority if they want to: they can evaluate him poorly, they can complain to the program director, or they can withdraw their attention in class. This acts as a force pushing the boat away from its destination, since he must pay attention to the satisfaction of the students to ensure he is not deauthorised.

Dan also has constraints imposed on him by the bigger system: the people that hold positions of authority over *him*. Some may be disappointed that the length of Dan's problem sets takes time away from other subjects. There may be those who expect him to teach in a certain way, perhaps more in line with the traditional method they were taught themselves. And there may be those who do not see Dan's passion for teaching as an important asset to the school, preferring to promote top-class researchers.

Even when we define it as a collection of vertical and horizontal bonds between teachers and students, there are still competing ideas of what the 'learning environment' might be. Harvard, in its widest sense, is one learning environment. It has numerous colleges, houses, departments and graduate schools all under the responsibility of its President. The Kennedy School is one such graduate school, led by its Dean. His job, too, is to strengthen and protect the learning environment within the school. Within the Kennedy School, the students are separated into programs led by Program Directors; these programs are comprised of one or two years of lectures, tutorials, seminars, problem sets and office hours.

The President, the Dean, the Program Directors, the lecturers, and the teaching teams are all trying to strengthen bonds to facilitate learning. By focusing on Dan's work and considering only the learning ecosystem under his purview, we are simplifying the wider system. On the question of why Dan's methods seem to be rare despite their success, we need to consider the question of incentives for different actors. Harvard maintains its reputation largely on the strength of its research. Students are attracted to Harvard as a place that pushes the barriers of knowledge, and will pay premium tuition fees for the privilege of studying there.

Students also have an interest in receiving top-quality teaching, but this is rarely the main driver for their choosing Harvard. The university has an incentive, therefore, to promote top-quality research ahead of top-quality teaching. They do a decent job of encouraging both, but a world-class researcher who is an average teacher is a much bigger draw for Harvard than a world-class teacher who is an average researcher.

Dan falls into the second category. He has research, and field experience, to his name: co-authors of his include the Nobel Prize winner Michael Kremer. But his passion is for teaching, not research. He has tried, in his time at the Kennedy School, to improve teaching at the school in broad terms: he is the chair of SLATE (Strengthening Learning And Teaching Excellence), the school's main initiative for improving teaching and learning. This has been challenging at times, as he himself admits. In a recent interview, after the coronavirus pandemic forced universities to move online, he said: "For the past nine years, I had been trying to get our school to do more – and better – online learning. I spent hundreds of hours, attended dozens of meetings, sent thousands of emails in these efforts, and had mixed success. While I would like to think that all these efforts helped the school be better prepared when we were forced to move online, it is fair to say that a virus of less than 1 mm in diameter has done more for online learning at the Kennedy School than all of those meetings and emails combined."

At a place like the Kennedy School, it is not just research that acts as a force pulling professors away from world-class teaching. Many professors spend time advising Presidents and Prime Ministers in their home countries: especially in times of crisis, this can lead to the quality of teaching falling down the priority list once again.

These constraints are part of the system, and mostly inevitable. For those with a passion for teaching, they are frustrating. But perhaps, given that research – and in the case of the Kennedy School, public service – is one of the end goals of teaching, we need to confront the challenge in a way that sees research and teaching not as forces pulling in opposite directions, but towards a common goal of the public good.

13

THE INVISIBILITY OF HEAT

O f all the invisible elements of a classroom environment, perhaps the most difficult to measure is the level of 'heat' in the room. Heat, for our purposes, is generated when cognitive work is done. The brain is, in some way, under pressure. This does not have to be directed towards learning: someone who is under stress because they are struggling to meet rent payments may be generating a huge amount of heat, and this could prevent them from learning.

Once again following Heifetz, we can think of the learning environment as a physical container[xxxiii], constructed from the vertical and horizontal bonds between the members of the classroom. The stronger the container, the more heat it can withstand: the stronger the learning environment, the more work will be done. This is why strengthening those bonds – for instance, by knowing everyone's names and backgrounds, or encouraging student-to-student interactions – is so important. It allows Dan to push the students harder, and hold their attention for longer.

In the analogy of the sailboat, heat is generated in the engine when the motor is turned on. This powers the boat forward towards its goal. However, the engine can only deal with a limited amount of fuel at once: too much could risk a fire. Every container has a limit as to what it can withstand, and even in Dan's classroom, where the bonds between members are strong, too much heat may be damaging. The invisible bonds need to be put under pressure for cognitive work to be done, but the level of pressure must be managed within acceptable levels.

Heifetz tackles the challenge of 'regulating the heat' by describing three zones for the temperature in the room. When the heat is too low, students are in the 'comfort zone'. They are not under stress, and likely not to be learning much. We saw earlier how 'superstar' students often end up answering all the questions in non-inclusive classrooms: this allows the others in the class to sit through the lecture in the comfort zone. As soon as the professor increases their wait time and avoids the superstar students, the heat in the room increases. This is what we meant earlier when describing silences as a 'subspace' in which learning can occur. The higher wait time is pushing a cognitive load onto the students.

Conversely, when the heat is too high, students enter the 'danger zone'. The stress level is too high for them to think clearly, and learning does not happen. There is too much heat in the furnaces, and too much pressure in the container, leading to a risk of explosion. Often the practical response to students entering the danger zone is to reduce the heat by taking out their phones. This drops them immediately down to the comfort zone, where they stay for the rest of the lecture.

In some classes, students will *start* the class in the 'danger zone' and the lecturer's immediate goal will be to lower the heat to get to a stage where learning can take place. A good example of this, as we mentioned in Chapter 5, was when the coronavirus pandemic took hold in the spring of 2020. For many students worried about loved ones, or returning home, or financial concerns, it was difficult to concentrate on the task at hand. Lecturers faced the challenge of attempting to reduce the tension while still maximising learning within the new constraints.

The 'learning zone' falls between the comfort zone and the danger zone. This is where students are forced to think, but not beyond their capacity. The sailboat makes good progress towards its goal. The teacher's challenge is to keep the classroom in the learning zone for as long as possible. Those old videogames come to mind where you need to navigate your helicopter through obstacles without letting it touch the top or the bottom of the screen. The learning achieved in any class will be proportional to the sum of the time spent in the learning zone across all the students.

An obvious challenge is that every student is different: the same material may push one student into the danger zone while keeping another in the comfort zone. As usual, the teacher will have to find the right balance across the class. In most classes, students' attention span will fall as the class continues, making it harder for the teacher to remain in the learning zone as their battery level starts to approach zero. This is the reason they keep TED talks to eighteen minutes. To manage these challenges, the teacher needs to be able firstly to take the temperature of the room, and then to be able to regulate the heat in response.

Figure 13.2: the learning zone

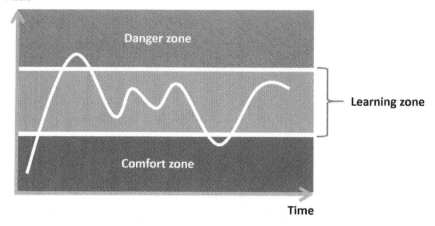

TAKING THE TEMPERATURE

We spoke about 'access to information' as one of the resources that comes with the teacher's authority: this included being able to take the temperature of the room by observation or poll. Confused faces, or poll results showing a lack of understanding, are a good indicator that the heat may be too high. Bored faces, and polls showing that everyone 'gets it', may indicate that the heat is too low.

I asked Dan in interviews how he takes the temperature of the class, and whether there are specific things that he looks for. He makes the point that it is important to distinguish between *mastery*, which he finds very difficult to gauge, and *engagement*, which is easier.

"How do I know if they're mastering the concept?", he asks. "I don't trust my ability to use non-verbal language to give me a sense of that."

He mentions the story from Chapter 12 of the time he thought the class had understood a concept, but a poll showed only 17% of them getting the answer right. "To assess understanding or mastery, I feel pretty strongly that I need that evidence. It could come from a poll, or a discussion, but I don't feel like I can look at them and just know. To assess engagement, there are non-verbals I look into. In general, prolonged silence worries me."

I ask him how his worries about silence can be reconciled with the fact that he uses silences so often as a tool in class, and his claim to the students in the first class that he is comfortable with silences.

"I like the distinction [from the monk Thomas Merton, in Chapter 6] between positive and negative silences. I don't worry about the silence after I ask a question and wait for hands to go up. I worry more if they don't seem to be engaged with the question. If there's processing going on then that's great, but if I'm interacting and don't see them interacting back, then I feel uncomfortable. If they're not asking questions, if there's no tension between the students and the material, then I'm not doing my job."

I ask him how he makes that distinction: how does he know, after asking a question, whether he is in a positive silence or a negative one?

"I'd like to think I know how to distinguish between the two by looking at the class and sensing the level of engagement, even if I can't sense whether they understand a concept. But I could be fooling myself: not everyone has to be experiencing the same kind of silence."

He continues: "I think the most difficult job as a teacher is putting yourself into the shoes of the students. There are lots of them. One might think, 'what a great debate we're having'. Another might think 'what's he trying to teach us?'"

He cites the example of the discussion way back in Class 2, when the doctor in the class criticised the debate about whether women in their forties should get mammograms as too simplistic. Some in the class thought that was a great discussion, helping to bring the content into the real world. Others, especially those just getting to grips with probability and Bayesian methods, saw it as a damaging distraction from learning the fundamentals.

Dan goes on to talk about assessing the emotional state of the class. He often interacts with the front row before the class starts, asking them about the mood in the class. "If they say they've just come back from a tough exam, then that feels like a different environment. If they've all been out until 4am the night before, then there's just no way the class is going to go well unless I acknowledge it."

Dan's approach to taking the temperature is a good example of the *lack* of magic in his approach. Some other professors may wonder how Dan can possibly gauge the level of understanding in the room and adjust as he does: the connection with his students seems impossible to recreate. But because he cannot observe the level of tension in the room, he seeks out data to tell him. The discussions with the front row before class, as well as the use of Poll Everywhere and Teachly, all give him data that help him to understand the invisible dynamics in class.

REGULATING THE HEAT

There are four key tools that teachers can use to raise the heat in class, with the goal being to move the class from the comfort zone to the learning zone. To raise the heat, they can speed up the flow of information, increase focus, create productive silences, and orchestrate conflict. Conversely, to lower the heat, they can slow down the flow of information, relax the class, give them answers, and avoid conflict.

Speed of information flow

Changing the speed of the information flow will change the cognitive load the students are asked to bear. Even though Dan plans for the length of time that each section will take, he treats the plan as a guideline. When polling suggests that the students understand, it may be worth forgoing a planned discussion in favour of moving onto another topic. The goal of any discussion is to bring the students into the learning zone: it needs to force them to think. If they get the concept already, they will remain in the comfort zone and learning will not take place. Conversely, if the polling shows that they are struggling to understand, the information flow may have to be slowed down. In these instances, Dan will spend more time taking questions, as we saw in Class 5. He will also spend more time on fundamentals than extension areas.

One challenge for the teacher in regulating the speed of information flow is that they may have a limited amount of material to get through. In most lectures, Dan prepares extension material in an appendix, which he presents in class only if he has time. Contingency planning allows him to remain focused on the students: without it, there is a risk that his own mind will be distracted by the task of rearranging timings in class.

Focus or relax the class

We have spoken about the importance of having a clear sense of purpose around maximising learning, and the challenge of recognising and mitigating the competing purposes of the class. There are times when enjoyment, or other competing purposes, will distract from learning: students may stay in the comfort zone as a result. Just as a soccer manager will shout at players to focus if they appear complacent, Dan can reassert the purpose of the class in an attempt to raise the heat. When we have heard him tell the students that his purpose is to maximise learning and not satisfaction, he is providing focus, and turning on the motor of the boat.

Done at the right time, an instruction to focus can get students back on the edge of their seats. The challenge for the teacher is to identify whether the students' lack of learning is because they are in the comfort zone or the danger zone. Bringing focus to students who are complacent will be a different task to bringing focus to those who are anxious. When students are in the danger zone and their capacity for learning is low, Dan sometimes opts to take a break. He tells the class to stand up and stretch for a minute or so, hoping that a bit of physical looseness will ease the pressure on the brain. When lots of students are in the danger zone, battery levels across the class are low, and attention spans are short. Bringing the heat down also serves to charge the batteries, increase attention spans, and return the class to focus for the rest of the lecture.

One student sees this ability to raise and lower the heat, while still creating a psychologically safe learning environment, as the key reason for Dan's success:

"Genuinely the best teacher I have seen in any course ever. Even as someone with heavy use of statistics in the past, I have massively improved my understanding, and his ability to push everyone just the right amount so that they feel safe and comfortable, yet perform far beyond their ability, is a masterclass in how to keep students interested."

The revealing part of this comment is how the student links the three ideas of being 'pushed', feeling 'safe', and staying 'interested'. They are saying that the best way to keep students engaged is to raise the heat by 'just the right amount' so as not to feel unsafe: this is analogous with our concept of the learning zone.

Use humour to release tension

We discussed in Chapter 5 how Dan's use of humour helps to support the learning of the class by taking a competing purpose – maximising enjoyment – and directing it towards the work that needs to be done. A more important use of humour is in releasing tension in the room. We spoke about how stretch breaks can recharge batteries at a time in the lecture when students have entered the 'danger zone'. But this can only reasonably be done once in a single class, whereas attention fluctuates by the minute. When the class laughs, brains switch off for a second and recalibrate. This helps to temporarily recharge battery levels by small amounts each time, and allows the class to pay attention to the next topic.

We defined battery levels in terms of authority: they are a measure of how much each member of the class is *authorised* by the others. Since maximising entertainment is a competing purpose for all students, to varying extents, humour will be one service they will hope for a professor to provide. Everyone wants to laugh from time to time. Given that we authorise people more when they provide the services we are looking for, lecturers who use humour will tend to build stronger bonds.

That said, a strong learning environment needs more than just strong bonds between its members. A comedian who knew nothing about statistics could develop these, but no learning would be done. The environment needs to be directed at the purpose of maximising learning: like any holding environment, its purpose is to facilitate adaptive work. If a professor over-uses humour to get students to like them, the strong bonds they create will be in vain.

Dan's use of humour is mostly well-directed, and tends to support the learning environment. The one area where he may sometimes miscalculate, as we mentioned back in Chapter 1 when talking about establishing norms, is in his reaction to late students: norm enforcement requires a *raising* of the heat when norms are broken, and humour serves the opposite purpose. This is partly his calculation: he believes that being dictatorial, in almost any scenario, is not worth the damaging effect on the invisible bonds within the class. However, it is also his nature: for better or worse, it is difficult to imagine Dan using the sort of firm voice required to lay down the law.

Create productive silences

We have already spoken about silences in two contexts: first, the way they can increase inclusiveness, and second, the way they can support psychological safety. A productive silence also raises the heat by giving work back to the students. We can think through our mental process when a difficult question is asked in class: our mind starts working, trying to make sense of it. While this is happening, we are in the learning zone. Without a silence, we get the answer from a superstar student, and we immediately fall back down to the comfort zone: think back to Mary Budd Rowe's research, which showed that the average response time in class was just one second. This was the whole time we spent in the learning zone, and perhaps we were not there at all if we knew the superstar student was going to give us the answer.

The teacher will often be able to observe the move from comfort to learning zone in the faces of the students. The silence creates a bundle of potential energy, a tension waiting for resolution. Psychologically, we are trained to seek resolution to tension: think of the catharsis of a match-winning goal, or the ending of a romantic movie, or the satisfying closing chords of a piece of classical music. This urge leads us to want to cut off the learning as soon as possible because we want the question to be answered.

This leads to a norm in which silences are uncomfortable: a norm so strong that we might even think the teacher is being rude by ignoring the first student to raise their hand. As well as normalising silence as we saw in the first class, Dan also tells them he will not always call on the first hand raised. This allows silences to be productive, as students do not stop thinking once the first hand is in the air. What seems like a simple change to teaching style can be hard to implement because of the need to override the urge for tension resolution.

ROBERT FRITZ: SEVEN GLASSES OF WATER

The American composer and film-maker Robert Fritz explored in detail the extension of tension and resolution from music to art and nature. In his book *Creating* (1991), he discusses an experiment called 'Seven Glasses of Water', in which he starts a workshop with seven empty glasses next to a pitcher of water[xxxiv]. Just before the day begins, he fills one glass with water. Before the morning break, he fills a second glass. On being asked what the glasses are for, he simply answers "that's a good question".

At the start of the afternoon session, he fills a third glass, leading some of the group to snicker. During the afternoon break, someone asks him if she can drink from one of the glasses, and he says "no". She says "oh!" and walks away with a knowing smile. Fritz wonders what it is she thinks she knows.

The remaining glasses are filled over the course of the rest of the day, leading to more and more questions from the participants. When the final glass is filled, there are claps and cheers. Fritz asks them all to write on a piece of paper what they think has been going on with the glasses of water. The answers vary from "he will play music on them" to "they symbolize the universe, filled with the purity of life". One has a complicated theory about the order of the glasses that were filled. Some think it's a type of ritual.

Having read the answers, he explains that it was a tension-resolution system. In discussions afterwards, one of the participants who thought he would play music on them says that she felt good when she 'worked it out'. Fritz points out that she did not really know what was going on, and yet she still felt better. She admits that creating her own 'resolution' made her feel more in control, and less likely to 'screw up'. Resolutions are an important psychological tool to make us feel reassured in this way, even if we are forced to invent them

As Fritz highlights, there is an overwhelming temptation to resolve tension because of our need for control. We like to be in the comfort zone. In class, it feels right to call on the first student with their hand up, and then it feels right to validate their answer. To create a productive silence, the teacher and the students need to let go of this desire for control. Doing so can be a liberating experience, and this is one of the reasons that students see Dan's class as such a breath of fresh air.

The opposite of 'creating productive silences', in this context, is to give students the answers. When this happens, the subspace for learning created by the silence is closed out, so Dan tries to prolong the tension as long as possible. However, when he assesses that the class is getting diminishing returns from the discussion, he will provide the answers the students are looking for. Sometimes he will do this himself; more often, he will call on a student he knows is likely to give him the answer he is looking for, as we have seen at various points in the book with Steven, Gabriela, and Rajiv. If students can learn effectively from each other, this is always preferable, partly because it preserves his own battery levels. By the time he intervenes, the students are often highly attentive to hear what he has to say: his battery has returned to full charge.

We can also see humour as the end of a tension-resolution system. This is not just true in the classroom, but in life in general. In tense situations we seek out opportunities to release the tension with humour. It is noticeable in Dan's classroom, and in many others, that there will often be laughter when a silence is broken after the students have struggled with a difficult concept. Laughing tells us "that's okay": we do not need to understand everything on the first go. In this sense humour is also part of psychological safety: if we are empowered to laugh about our struggles then we will not see them as a sign of failure. Tension is vital for learning, but in a psychologically safe environment, that tension is released in a way that does not demotivate the students.

14

IMPROVING THE CLASS: THE VALUE OF DATA

At the start of Class 5, we saw Dan recounting several years of his own self-evaluations on that lecture. The effect was to bring the students into the pedagogical struggle, and put responsibility on them for their own learning. After each class, he takes copious notes and also invites students and observers to give him as much feedback as possible. His reflections for the 2019 class run to ninety-four pages, and 32,000 words.

The effect is to give him a huge amount of behind-the-scenes data with which to improve the class on a continuous basis. We can add this to the data he gets from his teaching assistants, from the Teachly app, from the Poll Everywhere questions, and from the mid-course and end-of-course surveys. Dan is obsessed with data collection: one dream is to develop the course software further so that it can identify the mistakes each students make, compare them with past mistakes, and automatically find areas in which they are struggling. School teachers have done this for years, but the capacity is often not there for marking to be done in such depth in universities. Advances in automation and natural language processing may help.

The self-evaluations tell another version of the story outlined in this book. They highlight the internal tensions that Dan wrestles with in every class, always toying with the question of how to maximise learning. They are often inconclusive: he does not know the best approach, but still puts his struggles down on paper to read again next year. As we said in Chapter 5, the invisibility of learning makes progress difficult to measure. The best we can hope for is to get as much data as we can, and be ready to adapt and iterate, just like the Lewis and Clark explorers in 1803.

START OF TERM: UNCOVERING OUR BIASES

In the first class, with the Trump election example and the birthday game, we discussed Dan's decision to begin with a lot of behavioural psychology. The relationship with statistics, we said, was more important than understanding mathematical proofs, line by line. Dan's self-evaluation reveals some tension in this decision.

"Since there was no hard math in the class", he writes, "it might have given people the impression that this class was not quantitative at all. Maybe I need to frame better that this opening class is supposed to be more motivational and less quant-heavy than others – or perhaps actually solve the birthday problem."

Even though Dan's focus is always on statistics in the real world, the students will have to do some number-crunching as the course progresses, especially in the problem sets. He worries that students will get the wrong impression, and not steel themselves well enough for the demands of the course. At the same time, he wants the opening class to be accessible and fun.

Teachly data for this class also show that he under-called non-English speakers, and over-called white students, by a large margin in this class. There could be several reasons for this, including unconscious bias, natural variation, and greater confidence of English speakers at the start of the year. The data allow him to do something about it immediately, and try to be more balanced in the second class. The goal of Teachly is not to 'catch out' the teacher for their calling patterns: in any individual class, it's hard to stay completely balanced across gender, race, background and language. It just adds helpful data that the teacher can be aware of and react to if necessary.

The second class, on 'updating your priors' and mammograms, was the one where the class all answered the question on the updated likelihood of breast cancer given a positive mammogram test, and talked about whether Martin was more likely to be a finance professor or a physician.

"The thing that keeps bothering me about this class", he writes, "is that none of the polling questions require deep thinking, and the main work that students have to do individually by themselves is a simple calculation of Bayes' Rule. I would like to have time for a more substantive polling question."

We can see a similar tension to in the first class: underlying it is a fear that the first few classes are 'too easy'. He feels like he is not getting the students to do enough of the work in understanding how to 'be a Bayesian'. In this case I think he is too harsh on himself: the question about Martin, while a simple one, does require some deep thinking in terms of the concepts. The idea that Martin is more likely to be a physician than a finance professor, even though he invests in stocks and reads the Wall Street Journal, is mind-blowing to some students, who suddenly realise they have never evaluated their priors properly.

His fear may also be related to the time management of the class. He writes later: "I said that Bayesian thinking was an airport idea, but didn't give them much time to do work around this. Ideally, we would get there with ten minutes left, and spend some time thinking about how it applies to their own lives."

The reality is that Dan would love to be able to do more than he can pack into the 75-minute class. He wants to do more real-world application work, have a question requiring deeper thinking about Bayes' rule, and make it more demanding for the students. They already have mammograms, Martin, and Sally Clark's tragic case to consider; it is tough to get through everything.

Dan is pleased with Class 3, which focused on decision-making and 'resulting'. He writes: "I managed the discussion of regret vs. bad decision better than in past years. I said they should assume all the information in the problem was right, which may have ruled out explanations I got in previous years."

He is referring to the part when he asks the students whether they would regret not buying insurance on their iPhone if they broke it, and tells them that feeling regret is different from believing the decision to be wrong. His observation here is a good example of his self-evaluations from past classes acting as useful data. A common reason for discussions to get derailed is that students will go off on tangents by questioning the assumptions underlying the task. When this happens, it can be tough to regain control: Dan wants to move back to the topic at hand, but does not want to shut down the students' intellectual curiosity. If he can pre-empt this by asking them not to challenge the premise of the question, he can keep the discussion focused on the learning he wants to be done.

Class 4 is one of the highlights of the term, with the surprise guest appearance from Antonio in his World Bank office in Mexico City. Students reflect on this class fondly, but Dan's expectations of himself are high. He writes that the class went well, but he is frustrated with the fact that some students could not see his writing on the board. "I did not plan board work well", he writes, "and this is the same problem I had last year."

He is also disappointed that he was rushed for time at the end of the class. The precision in his time management becomes clear, as he expects himself to adapt minute-by-minute. He writes: "Managed time well enough to give Antonio 13 minutes. Should have given him 15."

The rush also meant that he did not have time to do a wrap-up of the concepts before handing over to Antonio. "It's hard to do it after Antonio speaks, as this takes time away from him. I should have talked about the trade-offs between leakage and undercoverage, and the role of statistics in *informing* the decision, not being the *only* factor. I should have said this at the outset as well."

The use of statistics as one input of many to inform real-world decisions, and not the sole input in some mathematical problem, is pervasive throughout the course. Even so, whenever Dan misses an opportunity to drill this point home, he considers how he can do a better job next time round.

MIDDLE OF TERM: INTERPRETING THE DATA

As we know, Class 5 on sampling distributions is the one Dan struggles with more than any others. We saw earlier that his summary for the class was that "this year was a setback", and that he needs to "rethink the class once again". In his self-identified areas for improvement, he goes into this in more detail. He is sceptical that the polling question, which was about taking a sample of 100 companies from the Fortune 1000 index and looking at the mean, was the right one.

"I managed time poorly, got stuck on the polling question, and never got to the Central Limit Theorem formally. I'm still not sure whether it's the right polling question, and a couple of people came up with tricky counter-examples. The second vote, after peer discussion, moved towards 'not enough information' instead of the intended 'correct' answer."

He also notes that two questions distracted the class from learning in the dice game: 'why is the sample size different with the roll of the dice?' and 'do we sample with or without replacement?'. As we noted in Chapter 5, the first question caused confusion because it blurred the important fact that we talk about a sampling distribution *of a given sample size*, e.g. 100, and in the dice game this size kept changing.

The replacement issue came about during the 'toy' example. In a normal sample, it is almost irrelevant whether you allow the same person, by chance, to be 'replaced' and sampled more than once because it will be lost amid the averaging effect of a high sample size.

If you sample just two people from a population of five, as Dan does in his toy example, replacement becomes more of a relevant question.

These questions are fidgety, and hold up the learning of the class. They support his course assistant's point that the 'leap' from the toy example to the Fortune 1000 polling question is bigger than he realised. But the interesting thing is the depth into which Dan goes to root out the problem. The class is not working well, and he needs to consider every possible way in which it might be being derailed.

Fortunately, the doom and gloom of Class 5 does not last too long, and he is pleased with the classes that follow. He notes in the next class: "I told a student who was defending an unpopular polling answer that I was applauding her bravery."

This is one way to exercise leadership, according to Heifetz. One of his pieces of advice[xxxv] for those leading from positions of authority is to "protect voices of leadership without authority". By protecting his student in that way, Dan was promoting psychological safety in the classroom and elevating an unpopular opinion. Not only is it *okay* to be wrong: it is actively encouraged. By expressing an idea, even a false one, the student helped the class to see the concept from different angles.

When the students reach hypothesis testing in Class 7, Dan is pleased with the addition of the image of the girl and the toothbrush, which was not in the previous year's class. "I strengthened this part by being clearer about what the null hypothesis is, and by asking them to speculate about how the alternative hypothesis might be true."

Asking the class whether it *might* be true that she brushed her teeth, even though the toothbrush was dry, was an effective way of getting across the logic of hypothesis testing. "It was great to make the point that we were not rejecting the null hypothesis because it would be *impossible* to observe the evidence that we did, but instead because it would be highly unlikely."

Moving to the class on cherry-picking, his overall summary, and advice to himself next year, is cautiously positive. "This class went better than last year. But don't get over-confident: the quality of this class has varied widely from year to year. Need to be ruthless with time management for this class to work."

We see that he struggled in previous years with managing time in this class, and he reflects that he was "more disciplined in the discussion" this time round. "I even had five minutes for them to write their takeaways at the end!" he notes, excitedly.

That said, there are still a couple of points that frustrated him. "I did not integrate the individual problem set responses on the Iranian elections piece, so I didn't look for a student who had identified cherry-picking." He does a lot of pre-class preparation, but cannot always cover everything in the students' problems sets. This comment highlights how important he sees the connection between the problem sets and the class.

"Three years ago", he continues, "a student came by and told me that he thought he was good at detecting cherry-picking, but that he had not seen it in this example. I should exploit this a bit more. Perhaps ask the class 'were you familiar with cherry-picking before this class?', and 'if so, reflect on why the Iranian election example stumped you?'"

Once again, Dan is keen to use polling questions to get as much data as possible from the students. Cherry-picking can be hard to spot even for those who have been taught about the concept. Perhaps students had prior beliefs, on seeing a question about Iranian elections, that the outcome would be to establish fraud, and were less 'on the lookout' for things that may work in favour of no fraud. If true, this would be an interesting example of confirmation bias for Dan to explore.

Whenever he criticises himself, he tries to come up with concrete changes he can make to improve next year. This is reflected in his comments on the causality class: "I think this went pretty well... part of the success in time management was that I got rid of some stuff at the beginning: the poll about correlation vs. causation, the Economist ad about econometrics, etc." These were based on reflections last year in which he had struggled for time in this class.

He does, however, see room for improvement in linking the airport idea of apples-to-apples with other parts of the class. "I thought the 'control group should mimic the counterfactual' idea was clear", he writes, "but I should have let them struggle a bit more with this issue." The word 'struggle' is revealing: while some might see the students struggling as an indicator of failure, Dan feels that in this class, they did not struggle enough. He cut off the discussion too early, bringing them down from the learning zone into the comfort zone in a way that failed to maximise learning.

However, it may be no coincidence that in doing so, his time management was significantly improved on last year: the trade-off between losing time and adding depth to the discussion is one *he* struggles with in every class. This is even more true in the class on RCTs. Having asked them for their opinions about RCTs in the problem set before the lecture, he is disappointed not to have time for a proper debate.

"Did not have *any* time for broader discussion of RCTs", he writes. "Main strategic mistake was that I spent too much time in the 'Did randomization work?' question. The discussion was great, but not as important as the big picture... there was lots of energy in the room around RCTs that was not channelled well because we didn't have time for a broad debate."[*]

Having chastised himself for not allowing time for this discussion, Dan looks at the data from Teachly to help him better understand. "I thought I had lectured too much", he writes, "but the number of comments (30) and unique participants (25) was on par with the average for the rest of the semester." This tells him that his problem was not that he spent too much time speaking, but that he structured the class sub-optimally and spent too much time discussing the Project STAR example.

He has a similar concern about the prediction classes: in the time available, it is sometimes difficult to have the big-picture discussion. In particular, he is sad not to have talked about some of the ethics of machine learning: "We barely touched on the ethical aspect. Last year, I showed a slide on a bank giving men higher credit limits than their wives as a result of a predictive algorithm discriminating against women. Should algorithms be forced not to include gender or race? These are important questions to grapple with in the era of Facebook."

In summer 2020, children in the United Kingdom were given grades for exams that had been cancelled due to the coronavirus pandemic.

---

[*] In the end, the class *did* have the debate he wanted. Led by Emily, they organised it themselves and invited Dan to moderate. It was a big success.

These grades had been based on a predictive algorithm, and there were ethical concerns about area-based discrimination: students with identical records were being predicted very different grades just because one was from a poor area and another from a rich area. The government was eventually forced to abandon the algorithm. Dan feels he missed out on an important discussion about these kinds of issues: given that the focus of the class is on policymaking, this would have been a great addition.

However, the prediction classes are still very new, and Dan is pleased overall with how they went. He is not seeking to be perfect, but does want to be able to improve each year. "I think this was a good class", he says of the first one. "I am glad I created space for it, and while execution could be improved, the building blocks are there for a better class next year."

He continues: "It was good to build the class from what they know. I liked starting with 'how would you use the tools you've learned to predict housing prices', as it meant we arrived organically at the issue of overfitting. Also, the scratch example is a very powerful one, and it grounds the distinction between causal and predictive in a nice way. It shows how sometimes prediction is enough."

The overwhelming sense from Dan's reflections is of a professor obsessed with recording as much data as he can to ensure his classes improve every year. The class has already benefited from 15 years' worth of iterations: hundreds of new ideas tried, many discarded, some retained. His hunger to get better is as strong as ever.

PART V: END OF TERM

15

BRINGING IT ALL TOGETHER

T he students are deep into November, and their first term at the Kennedy School is drawing to a close. The bright yellow leaves that covered John F. Kennedy Park at the start of the month have disappeared, giving way to a thick snow blanket. The Charles River, part-frozen, reflects a bright blue sky. Cambridge is cold at this time of year, but beautiful.

The problem sets are starting to get shorter as the course winds down, but the workload is not easing. Not only are they revising for final exams, but they must also work in groups to complete the 'final exercise': a piece of coursework that will see them apply the skills of the course towards a real-world example.

This year, there is a choice between three projects. The first is to work with a household dataset in Greece, exploring ways of reforming the benefit system to bring more people out of poverty. The second is a prediction exercise, in which they have to use their new machine learning skills to help a development organization to decide which villages in India to visit to implement a program to reduce school drop-outs. The third is for the macroeconomists in the class, working with Mexican government data to come up with a growth strategy for one of the states. All of the exercises draw heavily on the content of the course, but also require significant research for groups to familiarise themselves with the issues specific to their situation.

For many, the final project will become an obsession. Many of their other courses will be heavy on theory in this first year, and students tend to relish the challenge of a real-world example, even if it means more work. Despite warnings from Dan, they will spend their evenings, nights and weekends running simulations and trying different regressions. One member of the winning team on the Greek benefits project, on seeing the task, remembers how her eyes lit up. "*This*", she told a friend, "is why we're here."

The Harvard network adds an extra motivation for them: the winning teams will be introduced to the people working on these issues for real. The Greece team will meet with the World Bank manager responsible for the benefits reform proposals, the India team will meet with the development organization, and the Mexico team will *go* to Mexico for two weeks in January to work with the government to implement their growth strategy. Dan has a natural advantage in working at the Kennedy School that he can pull together all of Harvard's resources to motivate his students: the line between his class and the 'real world' is deliberately thin, but for the final exercise it disappears completely.

Of the twenty-seven classes in the course, only three remain. At the start of Class 25, Dan puts on the screen a tweet from a student named Sachin, who went around the Kennedy School in the early hours of the morning and took pictures of smiling final project groups.

No cherry-picking evidence here – the only people at @Kennedy_School on Saturday night are MPA/IDs working on a statistics final project. Hope you're proud, @danmlevy! #API209

"I want to say that this tweet warmed my heart", Dan says, to laughter from the class. "I know you have been working really, really hard on the final exercise, and I just wanted to tell you that I am thrilled about that: in fact, when I got this tweet, I even toyed with the idea of driving to the Kennedy School just to say hello to you all. It would not have been a popular move at home, but I did think about it."

Aside from the hard work, a few things are revealing about this tweet. The first is the fact that phrases like 'cherry picking', which are airport ideas in the class, are increasingly found in the everyday language of the class participants. It is a good sign of an idea getting penetration in the class when people start using it in sentences outside of classwork. The second is the desire to make Dan proud: the students are working hard because they want to, not because they have to. This suggests very strong vertical bonds and a successful 'contract': the students want to give something back to Dan, and make him feel like the effort he puts into the course is worth it. This is fostered in part by the recognition Dan gives his students, as we see from his supportive words when the tweet is on the screen. Thirdly, the tweet includes pictures of lots of students across many different project groups, indicating strong horizontal bonds within the class. The final project, just like the rest of the class, is a collective experience.

Aggregating evidence: women in Bangladesh

We are near the end of the course, but there is a contradiction to overcome. Right at the start, with the example of Trump's election, Dan told the students that "certainty is an illusion". This was an airport idea for the class, and it formed the basis of Bayesian updating. Students were discouraged from thinking about outcomes in binary ways, and encouraged instead to think probabilistically. Rather than thinking "Trump *will* be elected" or "Trump *won't* be elected", they should be Bayesians: start with a probability that reflects your prior beliefs about the election, then update that probability when new information comes in. The brain likes to deal in binary terms, but a good Bayesian overcomes that impulse and makes decisions based on probabilities.

However, this is at odds with the approach they took when they learned about hypothesis testing. They were taught to assume a null hypothesis is true, and only to reject it if the evidence suggested that the original assumption was highly unlikely. They ended up with a binary classification: either they rejected the null hypothesis, or they failed to reject it. Most of the time, researchers are hoping that they will be able to reject it, as this means that they found a statistically significant impact. A p-value of 0.049 is music to their ears, but a p-value of 0.051, just above the significance threshold, means failure. The two are virtually the same: the first is not a better experiment just because it happened to fall below an arbitrary threshold.

If we are being Bayesian, we should update our priors similarly for each: and yet, in most of academia, the p-value of 0.049 may mean a successful paper, while the p-value of 0.051 is consigned to the dustbin. While Dan was critical of this approach earlier in the semester, Class 25 is an attempt to resolve this contradiction, and prevent the students from falling into the researcher's trap of being obsessed with statistical significance*.

In the pre-class problem set, Dan left space for questions from students. He puts one of them on the screen from Vicente, a student from Chile, along with his photo – just as he did in the first class with Juliana, and many times since. As always, the photo gets smiles and laughter from the class. He asks Vicente to read his question.

"Is there a systematic way to address all of the evidence we have?", Vicente begins. "What should we look for when we have evidence across a pile of papers? How can we be sure that we are not cherry-picking when we do so?"

The question hints at the underlying tension between research papers and 'being a Bayesian'. Dan announces: "Today's class is about aggregating evidence, and it's devoted to helping to answer Vicente's question." He looks at Vicente, who smiles. He has been one of the quieter members of the classroom, and Dan is helping to elevate his voice.

"I'm going to present my attempt at being systematic", Dan continues. "Rajiv, in the pre-class work, mentioned that he has another method,

---

* This tension is part of a wider debate between the Bayesian and 'frequentist' schools of statistical inference. Frequentist inference, of which hypothesis testing is a part, does not seek to assign a probability to a hypothesis being true: it is either true or false, and the goal is to find out which. Arguably, this leads to a more objective approach than Bayesian methods, which rely on your 'prior' beliefs. A Bayesian might counter that if you have prior beliefs, even when they are subjective, you will make better decisions by using them.

so if you don't like mine, then go and have coffee with him." The class laughs, looking at Rajiv, and Dan pauses before moving onto the case study he has planned for them. As he often does, Dan is recognising experience in the room.

In the problem set before the class, the students were asked to take an initial look at a proposed experiment, and to do no more than five minutes of online research about the topic. They imagine that they are in Bangladesh, looking to improve healthcare for pregnant women and infant children. They want more pregnant women to visit health centres for check-ups, and as an incentive to do so, they propose to pay them a cash reward when they visit.

"Okay, here's the process", he begins. "We'll state our prior beliefs, find the evidence, aggregate the evidence, then update our priors. The process is an extended form of Bayesian updating, which as I hope you remember, is an airport idea of the class."

He takes them back to the beginning of term. "Back in Class 2, we saw that when a woman went to get a mammogram, there was an initial probability of having cancer, and then she got new information from the mammogram, and that gave her an updated probability. The idea of this framework is the same, but we're going to be systematic about the process of setting and updating our priors."

In the problem set, the students were asked to state their beliefs about whether the cash reward would affect the percentage of pregnant women who visit, the percentage of children who get vaccinated, and the height of the children, filling out the following table by placing an 'X' in each row:

Figure 15.1: the percentage of students who put an 'X' in each cell of the table, denoting their prior beliefs about the effectiveness of cash rewards for pregnant women visiting health centres

	Prior beliefs			
Outcome	Negative effect	No effect	Small positive effect	Large positive effect
% pregnant women who visit	0%	4%	35%	61%
% children who get vaccinated	0%	21%	59%	20%
Height of children	0%	50%	46%	4%

Dan asks for volunteers to explain how they set up their prior beliefs. Monica raises her hand: "I remembered a similar program in Mexico: there were loads of complaints because women lived too far away to go to the health clinics. So I put 'small positive' for each one because I didn't think there would be a big effect."

Dan nods: "So your priors were informed by knowledge of a similar program in a different setting, which made you less optimistic. Great – anyone else?"

Ling goes next. "I thought in terms of short-term mechanisms versus long-term ones. It makes sense that cash rewards would lead women to visit more, but it's not clear whether that translates into more vaccinations for kids, or taller children. I put 'large, small, none' in the three rows because I think the effect gets smaller as you go further away in the causal chain."

Dan praises her reasoning, and moves onto the evidence collection phase of the process. "You've established your priors", he says. "The next step is to find evidence. Ask yourself what evidence exists, and what evidence *could* exist, to help you answer the key questions."

Silvia, from Argentina, says she wants to know more about the figures before the program starts: what is the baseline from which they are looking to improve?

"That's a great question", says Dan, "and we're going to look at that in 43 minutes' time." The class laughs: this technique of deferring a question until later in the class, when he has planned to talk about it, is one Dan uses throughout the course. Students are often impressed when he gives exact figures: he is showing off a little, but he can do so because he knows, to the minute, when he expects each section of the class to begin.

Monique wants to know more about the transport links between rural areas and health centres; Cristina wants to know how exactly the cash reward works. Jose Luis, from Nicaragua, says it would be good to run an RCT. Emma, from the US, wants to do more qualitative work, asking the women what their constraints are. Diana, from Peru, is interested in how similar the women receiving money are to each other: high variance may mean it is harder to find an impact.

"These are all important questions", says Dan, "and in the real world you'd have a bit more time to consider them. Today, you've only got ten minutes."

He asks them to turn to the back of their hand-outs, where they have a collection of evidence from various similar studies done around the world. Although student engagement is high given the level of interest in the topic, energy levels are lower towards the end of term, and it is no coincidence that Dan has organised a class with a ten-minute group exercise during which they can recharge their batteries. In his reflections, he writes, "This was the right class type, given that they are in the midst of the final exercise." Once the time is up, he asks them to record their *updated* beliefs in the same table, and compares the two on the screen.

Figure 15.2: updated beliefs of students about the effectiveness of the cash reward program in Bangladesh

Outcome	Updated beliefs (prior beliefs in brackets)			
	Negative effect	No effect	Small positive effect	Large positive effect
% pregnant women who visit	0% (0%)	↓ 0% (4%)	↑ 54% (35%)	↓ 46% (61%)
% children who get vaccinated	↑ 2% (0%)	↑ 72% (21%)	↓ 22% (59%)	↓ 5% (20%)
Height of children	0% (0%)	↑ 82% (50%)	↓ 18% (46%)	0% (4%)

"Let's see where we are. It looks like there's been a change in your beliefs: mostly from optimistic to pessimistic. You think the effects are smaller than you thought before. Why did you change your minds?"

Charlotte points out that even though most of the studies of similar programs found some effect, one of the ones that didn't was in Nepal, and this is the closest geographically to Bangladesh. Federico points out that many of the other studies were done in countries richer than Bangladesh, where government capacity to implement the program successfully is likely to be higher. Dan agrees: "It's important to factor in external validity: how well does each study generalise to the population you're interested in?"

He continues: "The biggest change in your views was in the percentage of children getting vaccinated: does someone want to suggest why?"

Teresa, from Peru, points out that the data showed that 94% of children in Bangladesh are already getting vaccinated before the program, and this is higher than in the studies done elsewhere. Since there is less potential for improvement when baseline figures are high, this made her less optimistic about the experiment succeeding. Dan smiles, checking his watch, and looks at Silvia: "I said we'd get to this in 43 minutes, and we did it in 41." The class laughs again at his precision.

More polling shows that 92% of the class changed their prior beliefs when they looked at the new evidence, and 87% of the class changed their level of confidence in their beliefs. Dan is pleased with this result, as it shows that the class is actively engaged in Bayesian updating. However, he thinks he could have got the message across more vividly. He writes in his reflections:

"Last year, I asked for people to raise their hands if they had changed their priors. Almost everyone did. This was a really neat way to end the class, and it emphasised the message. This year, I did it with a poll, but I think the raise of hands was more powerful. More importantly, I forgot to reinforce the message verbally that it is okay to change priors."

It is time for him to wrap up, and send the class on their way to various Thanksgiving celebrations. "This was an artificial exercise", he tells the class, "but I hope that I've responded to Vicente's question. The big picture takeaway is that there will never be a study that perfectly informs the decision you're trying to make. You will have to base your response on imperfect evidence. State your priors, find the evidence, update your priors. Writing your priors in the first place is helpful because it can de-bias what you think about the evidence."

The class is coming full-circle: we are back to talking about Bayesian updating, decision-making, and cognitive biases, as we did at the start of term. Ten weeks later, though, the students have many more tools at their disposal to analyse evidence, and to update their priors accordingly.

16

LOOKING BACK AND LOOKING AHEAD

In the final problem set, Dan asks the students to rate every problem set question they answered across the whole course, and now he plans to show them the result. He has noted in past reflections that with exams approaching, it is difficult to raise energy levels in this penultimate class. To try to address this, he asks for a drumroll before the results come up on screen. The students are happy to oblige: as we have seen, they will always be engaged when they are part of the data-gathering process.

On the screen, Dan puts up the average ratings for every single problem set question. He also shows the confidence intervals around the averages so that people have a measure of the variance of the responses. The top-rated questions are the ones about elections in Iran, the one about HIV testing and Bayes' Theorem, and the case study about pensions in Mexico. The bottom-rated question, by some distance, is the 'Nerdy Friend' question from halfway through the course, when he asked the students to summarise what they had learned so far to a friend who was smart but not well-versed in statistics.

"I want to spend some time talking about the Nerdy Friend question", he begins. "It's important to me, it's dear to my heart, and every year it's rated the lowest question." The class laughs heartily: just as he did in the midterm feedback when he acknowledged the students were finding the problem sets too long, just this simple recognition is enough to get the class on his side.

"I hope you know I take feedback into account and I do it very seriously: I just want to explain why I have ignored the feedback on this question. Although it has a lower average than the others, it also has a higher spread: some of you rated it as the most effective, and others as the most ineffective." He puts up a comparison between two students' answers.

Figure 16.1: two contrasting student views of the 'Nerdy Friend' question

Effective	Ineffective
"The question where we write an email to a 'Nerdy Friend' was the most effective question in helping me understand the concepts of the course. In order to explain things in a simple manner, you really have to understand the concepts."	"I really suffered answering the 'Nerdy Friend' question on a sunny Sunday afternoon. In my opinion, it was not that useful, and pretty tedious."

There is some laughter at the 'tedious' comment, but also a recognition that the Nerdy Friend question is the most helpful for those students who have struggled with the first half of the course. Those at the top end may find this revisiting of ideas tedious, but Dan's priority is to help those at the bottom catch up. More subtly, it also reinforces the norm that the stronger students feel responsible for the learning of the rest of the class: this is crucial for the 'team learning' framework we described in Part II. They come to accept that their goal is not to maximise their individual learning, but to maximise the learning of the class.

"Now I want to play you a video", Dan continues. "I could spend the next five minutes telling you how important it is to communicate with a policymaker who's smart but not well-versed in statistics, but I want to try a different tact."

There is a sense of anticipation in the room. As any high-school teacher knows, the students' hearts leap when they hear they get to watch a video near the end of term. With energy levels low, this second-to-last class is the perfect time for it.

"The video is a French movie that explains why this is important. In the clip, there is a town in France which has run out of water, and they've called a rural expert to advise what to do."

The clip, from the movie *Manon of the Spring*, shows a hapless expert attempting to explain to a rowdy town hall that "the source of the fissure is not diaclastic; it's a Vauclusian resurgence. The water bed between two impervious strata levelled out and the water issued forth upon the upper stratum..."

"Just let him turn on the water – he can explain later!" cries out someone in the audience.

"It could take two days, maybe two years", the expert replies.

"Maybe a hundred years", says one of the council members to another, rolling his eyes.

"That can't be ruled out", nods the expert, missing the sarcasm.

"To hell with your theories! What can you do for us right now?" shouts the mayor, rising from his chair.

"We suggest you farm some land elsewhere", shrugs the expert, before being chased out of the room by a mob of angry townspeople.

Dan closes the video. "I hope that this image spoke a thousand words: *I never want to see you be that rural expert.* I want you to communicate in a way that's clear, and understandable by those who aren't as technical as you are."

He moves on to the next question in the survey. He asked students about the airport ideas of the course: "What are three ideas that you will remember five years from now?" Up on the screen is a word-cloud of their responses.

Cherry-picking is the most prominent, and also clearly visible are the airport ideas of Bayesian updating, hypothesis testing, and statistical significance. "You did a great job picking out the key ideas", says Dan. "There was one important idea that didn't receive as much attention, and I want to bring it back with one picture that I hope will be ingrained in your minds five years from now. That's the idea of the *counterfactual.* Finding a control group that mimics the counterfactual is crucial for finding causality."

Figure 16.2: student-generated word cloud of airport ideas from the class

He puts up the picture of Virat as a rock star from halfway through the course. The students laugh at seeing it once again. The word cloud is a clever way for Dan to gather data on which of the airport ideas have had the most penetration, and which need reinforcing.

He then asks the students to talk to each other about how they could have applied lessons from the course in the past, or how they might apply them over the next year. He gets the students to do this exercise standing up: "My notes last year said it was hard to get energy at this point", he says, "but we need to do everything we can to stay awake. I know there are a lot of things weighing on you, but for the next 45 minutes you will defy your body and your mind, and bring all your energy to it."

One student begins by saying how the course has taught her to think probabilistically. "We talked about the probability section at the start: the probability of having a disease, and the probability of the test being right. Before coming to this course, I felt very emotionally about this topic, and I was quite emotional when I was doing this problem set. I thought it was unfair to think about not encouraging certain people to do HIV tests."

She pauses, and then continues. "I was thinking about real-life events that happened before coming to school. But we learned that even if you have access to these tests, there are reasons why not everyone should take the test. I started thinking from a rational, not an emotional, perspective, and translating it into economic policy."

Dan thanks her for her comment, which was a heartfelt one. When emotions are involved, which they will often be in both our personal and professional lives, it can be harder to think probabilistically. That said, emotions also give us the passion we need to make change, so it is a fine balance we must strike between managing them and channelling them.

Ankit gets the class laughing by saying that when he bought his new iPhone, he did not buy insurance, after the class on making decisions. Dan smiles, and says, "I will be trembling for the next two years to receive an email from you telling me you lost a load of money and it's all my fault", which gets another laugh from the class.

Pedro, from Colombia, has been fairly quiet throughout the semester, but he speaks up now. "It's a funny example", he says, and hesitates. "I was going through a bit of uncertainty in my life, and there was this girl I liked."

Dan raises his eyebrows, and the class is on the edge of their seats.

"Since statistics is the science of uncertainty", he continues, "I thought maybe I've learned the tools I need to figure this one out."

"Pedro – you can pause at any time", Dan says, making sure that he is not feeling under any pressure to reveal anything personal. This also gets a laugh from the class.

"I was getting some mixed signals: some positive, and some negative. I wasn't sure how to interpret them. So I did some hypothesis testing: I started with the null hypothesis that she likes me, and said, if this is true, how likely is it for her to be sending the negative signals?"

He pauses: the class laughs and then goes silent in anticipation.

"It gave me a basic framework to conclude that she was not interested", he says, to pitying noises from the class. "But the lessons we learned here, I applied to my life, and it helped me to figure it out."

The class gives him a round of applause, in support of their friend. Dan smiles, and says, "I just got goosebumps! That was *magic*."

Silvia speaks next, and talks about how the phrase "a bad result does not necessarily mean a bad decision" was influential to her. "So if Ankit breaks his iPhone, he might have made the right decision even if the outcome was a bad one."

"That's one aspect of it", agrees Dan. "Another is that if something has a 20% chance of happening, it will happen one out of every five times. So you shouldn't say 'it happened, the forecasters were wrong'. If the weather forecast says there's a 20% chance of rain, you might want to bring an umbrella: it will happen sometime." He is circling all the way back to the first class.

Juan talks about cherry-picking: "Monica and I were talking about how statistics is an ethical endeavour: when conducting the analysis, you might be very tempted by cherry-picking. At times we may have sinned a bit in picking the evidence to fit the narrative. You need a certain level of ethical commitment to the truth."

Lina, from Germany, adds to Juan's comment: "Also, if we communicate information to other people, we'll often be in a position of trust. They will probably believe us, so we should be careful about things like cherry-picking before we do so."

Fernanda goes next, and she has a story to tell. "My boyfriend is a microeconomist, and I remember seeing some kind of Bayesian thing in one of his game theory problems, and it traumatised me so much that I thought it was really complicated and I hated Bayes. So when you said 'be a Bayesian', I was thinking, 'I don't want to be a Bayesian!'."

The class laughs, and Fernanda continues. "Now I feel much more comfortable with all these concepts. I would always confuse conditional probabilities and he would say 'you're thinking the wrong way'. This wasn't intuitive, but after the first problem sets it became much clearer to me. I think I can have a better dialogue with him now!"

Dan looks amazed, and laughs. There is a videographer in the front, filming this class as part of an education series at Harvard. "I think he's going to think it's a course about personal relationships", he says. "I promise, it's a course about statistics."

The way that students start to open up about their emotions and their personal lives in this penultimate class is an interesting phenomenon. It comes from having a very high level of psychological safety, combined with the fact that time is running out: the learning environment will soon disappear, and they want to take advantage of the safe space while it still exists.

If psychological safety is not as high, students will fear making connections between the course and their personal lives, for fear either of losing authority, being mocked, or being told that their personal reflections are not relevant in the classroom. The effect of this is profound: only in classrooms with high psychological safety will students make strong links between the content of the class and their personal lives.

The great American writer Maya Angelou famously said the following:

> *"People will forget what you said, people will forget what you did, but people will never forget how you made them feel."*

Since our personal lives are what make us *feel*, and psychological safety is what allows us to make the connections between the classroom and our lives, we have a clear link between psychological safety and memory of ideas. Five years after finishing the Kennedy School, the students will remember how they *felt* in Dan's class. That feeling is a strong base from which to conjure up the airport ideas of the class. Since they spend much of the class time applying those ideas to the real world, and strengthening the relevant neural pathways, that feeling will also lead them to answers when faced with real-world problems years later. Other classes with 'key takeaways' may be less successful even if those takeaways are presented clearly, because the students will not have the same feeling to latch onto when recalling those ideas five years later.

THE FINAL CLASS

On the last day of the course, the whole class is led by the students. One group from each of the three final project teams, chosen by Dan, presents their findings to the rest of the class. Special guests from the rest of the Kennedy School are in attendance, including the world-famous economist Dani Rodrik. The students do not know who will present until they are in the classroom, and Dan announces the presenting teams with a drumroll on the screen. Before he does so, he shows the first slide from all of the presentations set to the overture from the opera Carmen: an energetic piece of music to get everyone going. Having all of the presentations on screen also gives recognition to the students, some of whom might be disappointed not to be selected to present.

Before the presentations, Dan also gives each student a blue sheet on which every 'airport idea' from the class is summarised. If they end up losing everything else from the class, he says, he wants them to keep that blue sheet of paper so that five years from now, when they meet him at an airport, they will be able to recall the key ideas.

It is fitting that the final class is student-led, as it shows the final transfer of responsibility from Dan to the students. Throughout the course, Dan has encouraged the students to take responsibility for their own learning, and this is what has helped to build the invisible bonds in the classroom. Having Dani Rodrik in the room will lower the psychological safety of the classroom: students will want to impress him, so the fear of failure increases. But this too is deliberate: the learning environment is about to disappear, and the students will be using these concepts in the wider world outside of the strong container of the classroom. Having built that container, Dan now dismantles it in front of them.

The students present well. As Dan notes in his reflections: "There is no better way to end than having students present their work. And they did an amazing job. I was simply blown away. And I think Dani was too!"

With the presentations done, Dan gives the students fifteen minutes to fill out the course evaluations, during which he steps out of the room. One reason for this is for the students not to feel under any pressure as they write the evaluations, but there is another psychological reason: without Dan, only horizontal bonds remain. For those fifteen minutes, the students are left with just the container they built, and the blue sheet of airport ideas.

Once the time is up, Dan re-enters the classroom for the final time.

"We have five minutes left", he begins. "Let me say some final words. First, I want to give a big round of thanks to our teaching team. The course would not work without them, and I want to ask not only for a collective applause, but also for them to come down here to be part of the final celebration."

Dan centres the class's attention on the teaching team, who have sat throughout the course at the side of the room. Bringing them to the front and centre of the class gives them recognition, and highlights to the rest of the class the role of the teaching team in strengthening the bonds in the learning environment. His last speech is one which includes the word 'feel' eight times, magnifying the emotions of the classroom.

"My final words of thanks", he continues, "are to you. As I left this classroom on Tuesday, I felt incredibly lucky. I had read your problem set answers to the things you had learned in this class. As I was hearing Pedro talk about how he used this course in his personal relationships, and how Fernanda used it to decide how she can have a better relationship, I felt truly blessed to be your instructor. I want to say a big word of thanks."

He pauses for effect. "I also want to give you a word of advice, as this is our last time together in a classroom like this. This is based on how I felt this week. As you make your career and life decisions, my main piece of advice is that whatever you pursue, you pursue with passion. Your life will be much better as a result, you will have greater impact in the world, and you will just be happier in your life. I feel very, very lucky to be in a profession where I feel that. My mantra to you, the one *I* have used in making career decisions, is 'will this job make me want to wake up at five in the morning to do the work?'. That's how I feel, and it's exactly how I felt every day coming to this classroom. So, a word of thanks to you on that."

He pauses again. Some students have tears in their eyes: they are experiencing a loss, and they are also aware of the loss that Dan is feeling. In a couple of minutes, the safe space they have enjoyed in this class will evaporate.

"My last word is that you will all go out after graduation and have great things happen to you in the world. I will be the first to tell you that correlation is not causation, so I won't take any credit for the things that happen. But when they do, please share them with me. I'll be cheering for you, I'll be rooting for you, and I feel very fortunate to be in a position to tell you that, and I will be very happy for you when those things happen."

The classroom is smiling, even if through tears: it is powerful to be told that a person is rooting for you. No matter how strong and independent we want to be, most of us crave support and validation on a deep level, and his speech helps them to realise that even if the learning environment will be dismantled, the invisible bonds built up between the members of the classroom will not evaporate when they leave for the last time.

"At the same time, if you ever have setbacks, or need a sounding board, or a crying shoulder, you have lifetime rights to my office hours. I want to say, from the bottom of my heard, thank you so much for making this semester an incredible experience for me. Thank you!"

Dan gets a standing ovation from the students, which does not stop when the clock ticks past the 2.30 mark that signifies the end of the course. They take pictures, exchange stories, and finally pack up their things and leave. Even as the physical space of the classroom disappears, the invisible bonds they have formed, and the invisible learning they have done, will stay with them. Some of them, five years from now, will meet Dan at an airport. And they will know exactly what to say.

Postscript

This book has been deliberately told from the student's perspective. In the spirit of getting across the way Dan's class made students feel, I wanted to end with a message sent by Juliana to Dan in January 2020, a month after the class ended, and quoted with her permission. She was the first student to have her photo and words up on the screen, and described how it made her feel like he cared. Given her passion for education, it is no surprise that Dan's class would have had an inspirational effect.

Her message echoes many of the themes of this book. She starts by echoing some of the other students' comments on the humility they have discovered from being in the class.

"Dear Dan,

I always finish my year looking at the positive events that have happened in my life – I know, I have this positive bias issue. I have learnt a lot, starting by being less intellectually arrogant and being open to say 'I don't know'. This was especially true with Statistics, the course I was the most confident about, and ended up having a way worse result than expected (which has taught me a lot in the end!)."

The fact that she still feels so positively despite a disappointing exam result reflects the work Dan does to avoid test scores being the key motivation for learning. Juliana then talks about how it felt to be included in class, both by Dan and by her fellow students, after previous experiences at universities in which this had not been the case:

"I am also very grateful to my new friends, who welcomed me and supported me, as well as the faculty members. And here I would also like to add a note to you. I don't think I told you, but the day I went to your office hours to talk about pedagogy was the first time I went to office hours with a professor in my life. I was feeling a bit anxious, but I was so curious to ask you those questions that I took the courage. And it was easy and great: I felt like I was talking to a friend.

In my home university, speaking in class was seen as a bad intervention, and I had two teachers that used to expose the girls who participated, making fun of us. This fear was still inside me and I have to admit that in the beginning of the semester, I was terrified to participate in your class. I could have done a way better job participating more, but thank you so much for making us all feel welcomed and in a safe environment."

As we explored in Chapter 7, psychological safety has a profound effect on the learning environment. For students like Juliana with bad past experiences, the right support and nudging can reverse fears of participation. Having a professor who cares about his students can be a cathartic experience that overturns years of difficulties in the classroom.

She then goes on to describe an experience she remembers vividly:

"I will never forget this scene that I captured in one of your classes: one of our classmates was hesitant to answer your question directed to him. You got closer with your friendly look, and put your hand on his shoulders. I could see the tension being released: he smiled back and calmly answered. He got it wrong, but it didn't seem to have bothered him."

This is an excellent example of Dan's ability to manage the heat in the classroom, as we described in Chapter 13. Sensing a student about to enter the danger zone, he held him through that challenge and brought the tension back down to enable him to learn. The effect was not only to support the student involved, but to make the space even safer for everyone else: this was what made it so memorable for Juliana.

She is back in Brazil when she sends the message, having hosted a group of classmates over the Christmas holidays. She concludes by sending some pictures of the group celebrating the New Year:

"The welcoming environment helped us create a community and become closer outside the class. The proof is below: 19 of us in Rio de Janeiro for New Year's Eve. Magic! We were all thrown in the water, and some people are still wondering why they hadn't bought phone insurance. Bad outcomes and/or bad decisions?

I just want to say Happy New Year, Dan! I'm sending very positive vibes to you and your family directly from Brazil. See you soon!"

Juliana is one of many students to have been inspired by Dan's class, and it is remarkable that she cites the class itself as a key reason for the strength of their community. His classroom may have disappeared, but the learning environment he created was strong enough, in the end, to outlast it.

Acknowledgements from David

This was an ambitious project, which would have been impossible to achieve without a great deal of help. Dan Levy, the subject of the book, was generous with his time in offering twelve interviews across the space of four months. This is particularly notable given that he was himself writing a book, 'Teaching Effectively With Zoom', aimed at helping teachers through the COVID pandemic. In those interviews, he was honest and open, and encouraged challenges to his way of thinking. Hopefully, this attitude comes across in the book.

I want to thank the entire class of API-209 in 2019 for their contributions and ideas about the book. To preserve anonymity, I have not called them by their real names in the book: this is important to maintain the integrity of the safe space in class. I give particular thanks to those who sent in extra comments and thoughts about Dan's teaching, including some from my 2018 class, many of which have found their way into the book. Alex Domash, Sam Elghanayan, Ruth Huette, Julia Liniado, Vale Mendiola, and Bia Vasconcellos were especially helpful in sending through ideas, thoughts and advice.

The MPA/ID program at the Harvard Kennedy School, in which the vast majority of students in API-209 are studying, is a fantastic experience in which the bonds between the students grow strong very quickly. This is primarily due to Carol Finney, the organiser of the program, who makes everyone feel at home within the first couple of weeks, and Dani Rodrik, the program head, who combines his role as one of the world's top trade economists with one as a support and inspiration to all the students. Deb Hughes Hallett, who teaches the Math Camp at the start of the program, is another brilliant teacher who has helped thousands of students navigate the challenging transition into quantitative methods at Harvard. Dan's success is aided by the environment that they all create.

Much of the framework in the book takes inspiration from Ron Heifetz and the approach he developed in *Leadership Without Easy Answers*. I would like to thank him for this, as well as Kim Leary and Hugh O'Doherty, who taught it to me at Harvard and provided helpful guidance along the way. Rahel Dette and Kudzai Makomva, my coaches for their classes, also had great influence on me as a person, and helped me to understand the links between the leadership framework and the applications in the classroom. Matt Andrews's phenomenal *Getting Things Done* class, which connects many ideas from the field of leadership to the world of policymaking, was also influential. It does for leadership what Dan does for statistics.

The teaching team for API-209 was a joy to be part of. If you ever wondered how Dan manages to put so many moving pieces together into his class, his faculty assistant, Victoria Barnum, does a magical job with the logistics. Jose Ramon Morales Arilla, the head teaching fellow for the course, and Shiro Kuriwaki, who helped the students through their struggles with R, played a huge role in making the year a successful one. Sophie Gardiner, a great friend with a fantastic intuition for statistics and pedagogy, came up with suggestions every week to make the course better.

My parents, Fiona and Mike, encouraged me to pursue the dream of going to Harvard, and have done more for me than I could ever describe. My sister Jenny, the only teacher I know better than Dan, has been an excellent sounding board for the pedagogical ideas in the book. My brother Euan, a writer and critic, gave me incisive and honest feedback on how to write. My grandparents, Jane and John – the 'Gan and Boompa' to whom this book is dedicated – instilled in me the love of learning that was the fire underneath this book. Boompa passed away last year, but was still taking university courses until well into his eighties and remained a loving source of advice and wisdom until the end. To all of them I am forever grateful.

The final acknowledgement is to Catri Greppi, who as well as being a phenomenal coach in both the statistics and leadership fields, also happens to be my partner. This book was her idea, and it would not exist in any form without her. She is known, especially by Dan, for her honest feedback, and this has been a great asset in improving the book and coming up with new ideas. But more than anything, I want to thank her for believing, against all odds, that I could write it.

Acknowledgements from Dan

David's book is about API-209, the statistics course that students in the Masters in Public Administration in International Development (MPA/ID) program at the Harvard Kennedy School (HKS) are required to take in their first semester. I have taught this course continuously every year in the period 2004-2020, including teaching it online in 2020.

I would like to thank the many people who have allowed me to teach this course in this program for so many years. First is Dani Rodrik, a world-renowned economist, who was the faculty chair of the program when I was initially asked to teach the course in the fall of 2004 as a visitor. Despite my best efforts and many hours of work, the course did not go well that year. I received the worst course evaluations that I ever received for any course or teaching assignment before or since then. This was a very humbling experience where I learned a lot, but my students unfortunately did not. Given their experience, I suspect that students from that cohort will be shocked to hear that there is now a book written about the teaching of this course! I am grateful to these students who provided me with many insights that led to improvements in API-209 and my teaching more generally. Dani has also done many things to support my teaching in API-209 since then, including giving me a second chance to teach the course in the fall of 2005 and currently serving again as the faculty chair of the program.

The MPA/ID staff has been successful in managing a program that attracts spectacular students and where many of the faculty involved are delighted to be part of this tight-knit community. Carol Finney has been director of this program since its inception in 2000, and has been crucial to its success and to the creation of this wonderful community. I owe special gratitude to Carol for engaging in many helpful conversations with me and for the many actions she has taken over the years that have directly benefited API-209 and the student experience more broadly. Her staff, including Sarah Olia and Kevin Drumm, are very committed to the success of the MPA/ID program and have been extremely helpful to all of us involved in it.

The API-209 course is part of a set of courses that students have to take in their first year, and I would like to thank the faculty with which I most closely coordinated to teach the course over the years. All of them are remarkable and I feel very fortunate to be able to work with them and learn from them. Nolan Miller taught the microeconomics course for many years. We were classmates at Northwestern University, and he was instrumental in bringing me to the Kennedy School as a visitor in 2004. Nolan, one of the brightest people I have ever met and a spectacular teacher of a course that is notoriously hard to teach, was a wonderful mentor to me and taught me a lot about effective teaching. Deborah Hughes Hallett, known to be one of the most effective and thoughtful math educators in the world, has been a constant source of wisdom and insight for me. She teaches students the underlying math tools they need in the MPA/ID core courses, and I would not be able to teach API-209 the way I do if it were not for her and her teaching. Jeff Frankel, a world-renowned economist and advisor to many central banks and governments around the world, has taught the first-semester macroeconomics course in the MPA/ID program every year since the program started, and has consistently been helpful and insightful in our efforts to integrate our two courses so our students get a more integrated academic experience. Ricardo

Hausmann, a Venezuelan compatriot and a world-renowned economist who advises presidents and prime ministers around the world, has been very supportive of the final exercise in API-209 and has been a constant source of wisdom and inspiration for me and for our students. The legendary Lant Pritchett directed the MPA/ID program for several years and influenced not only how I teach but also how I think about several of the subjects in API-209. Many other faculty members who have taught in the MPA/ID program over the years have helped improve API-209 as a course and my teaching more broadly. They include but are not limited to Alberto Abadie, Matt Andrews, Jie Bai, Filipe Campante, Eliana Carranza, Rema Hanna, Rob Jensen, Adnan Khan, Asim Khwaja, Maciej Kotowski, Eduardo Levy-Yeyati, Rohini Pande, Carmen Reinhart, Federico Sturzenegger, Arvin Subramanian, Andrés Velasco, and Mike Walton.

Integral to the design and delivery of API-209 are the members of the teaching teams that do a lot of the work to make the course run the way it does. In many ways, they deserve most of the credit for the course and I feel very lucky to have worked with them, and forever grateful for their incredible dedication and extraordinary commitment to our students. I list them here in reverse chronological order: Casey Kearney, Nicole Carpentier, Ruth Huette, Racceb Taddesse, José Ramón Morales Arilla, Shiro Kuriwaki, David Franklin, Sophie Gardiner, Catrihel Greppi, Felipe Jordan, Malena Acuña, Joao Araujo, Michael Lopesciolo, Jason Goldrosen, Astrid Pineda, Erich Nussbaumer, Layla O'Kane, Marie-Pascale Grimon, James Fallon, Jovana Sljivancanin, Vincent Vanderputten, Teddy Svoronos, Ishani Desai, Emilio Hungria, Maria Schwarz, Benjamin Maturana, Daniel Garrote Sánchez, Kate Sturla, Sandra Naranjo, Sarah Oberst, Shanthi

Philips, Mahnaz Islam, Aditi Chokshi, Rafael Puyana, Paola Vargas, Avnish Gungadurdoss, Asim Jahangir, Snezhana Zlatinova, John Graves, Chiara Superti, Andrew Fraker, Alexander Hall, Michele Zini, Matt Blackwell, Laura Faden, Kit Rodolfa, Olga Romero, Onur Yildirim, Aaka Pande, Conchita Galdon, Stefan Jansen, Andrew Myburgh, Allison Comfort, Matthew Blakley, Marc Shotland, Nina Stochniol, Holger Kern; Chester Chua, Tim Bulman, Jason Stayanovich, Bryan Wagner, Justin Timbie, Scott Griffiths, JP Ortiz, Niall Keleher, Cecilia Mo, Fei Yu, Arturo Franco, Kari Hurt, and Alfonso Tolmos.

I am obsessed with using every minute with my students wisely, and invest a lot of time in ensuring that the logistics of delivering the course go as well as possible. To do this, I have also been very lucky to count on the support of many people over the years including Mary Jane Rose, Katie Naeve, Alex Kent, Mae Klinger, Tamara Tiska, Sarah Hamma, Evan Abramski, and Victoria Barnum.

Many people have influenced how I teach statistics. Teddy Svoronos, now a faculty colleague at HKS, was a teaching assistant for API-209 in 2013-2015. He not only did a stellar job in the course but also helped turn it into a truly blended course by helping create incredibly compelling online modules that completely changed the way I teach. He has influenced every aspect of my teaching, and particularly how I deploy technology in the service of learning. I suspect that many years from now if someone were to mention what was my biggest contribution to teaching at the Kennedy School, they will say it was helping the school identify and bring Teddy Svoronos to our faculty. Richard Zeckhauser, a brilliant scholar, a wonderful mentor, and a great friend, is the person that has most influenced how I think about statistics and about the world, and has greatly influenced the content of the course and how I teach it. At the Kennedy School, I have taught other statistics courses and have had wonderful partners who have influenced the way I teach and think about statistics. Two that deserve special mention are Jonathan Borck and John Friedman. We taught a course similar to API-209 together for several years, and both of them deepened my understanding of the subject, helped me become a better teacher, and have been a joy to collaborate with. Others that deserve special mention and my gratitude include Chris Barr, Herman Bennett, Mike Callen, Steve Cicala, Suzanne Cooper, David Deming, Rachel Deyette Werkema, Sue Dynarski, Josh Goodman, Erich Muehlegger, Kerrie Nelson, and José Carlos Rodríguez Pueblita. Finally, some colleagues outside the Kennedy School have had a positive influence in the way I teach statistics by providing me with great insights and being wonderful examples to follow. They include but are not limited to Joe Blitzstein, Andrew Ho, David Kane, Gary King and Xiao-Li Meng.

I am also grateful to many colleagues who have influenced my teaching perspectives, including but not limited to Bharat Anand, Arthur Applbaum, Chris Avery, Mary Jo Bane, Matt Baum, Erin Baumann, Bob Behn, Iris Bohnet, Derek Bok, Peter Bol, Josh Bookin, Jonathan Borck, Dana Born, Matt Bunn, Sebastian Bustos, Gonzalo Chavez, Piet Cohen, Anjani Datla, Jorrit de Jong, Akash Deep, Pinar Dogan, Jack Donahue, Erin Driver-Linn, Greg Duncan, David Eaves, David Ellwood, Doug Elmendorf, Mark Fagan, Maria Flanagan, Michael Fryar, Archon Fung, Marshall Ganz, Alan Garber, Patricia Garcia-Rios, Steve Goldsmith, Tony Gomez-Ibañez, Merilee Grindle, John Haigh, Kate Hamilton, Frank Hartmann, Ron Heifetz, Dave

Hirsh, Daniel Hojman, Jim Honan, Kessely Hong, Deb Iles, Anders Jensen, Doug Johnson, Tom Kane, Felipe Kast, Steve Kelman, Alex Keyssar, Adnan Khan, David King, Gary King, Mae Klinger, Steve Kosack, Michael Kremer, Robert Lawrence, Henry Lee, Dutch Leonard, Jennifer Lerner, Jeff Liebman, Dick Light, Horace Ling, Rob Lue, Erzo Luttmer, David Malan, Brian Mandell, Jane Mansbridge, Tarek Masoud, Janina Matuszeski, Quinton Mayne, Eric Mazur, Tim McCarthy, David Meyers, Ian Michelow, Matt Miller, Francisco Monaldi, Mark Moore, Juan Nagel, Angelica Natera, Tim O'Brien, Tom Patterson, Allison Pingree, Roger Porter, Samantha Power, Todd Rakoff, Fernando Reimers, Hannah Riley-Bowles, Juan Riveros, Chris Robert, Chris Robichaud, Todd Rogers, Lori Rogers-Stokes, Eric Rosenbach, Jay Rosengard, Soroush Saghafian, Tony Saich, Miguel Angel Santos, Jeff Seglin, Anna Shanley, Allison Shapira, Mark Shepard, Kathryn Sikkink, Judy Singer, Malcolm Sparrow, Rob Stavins, Guy Stuart, Federico Sturzenegger, Arvind Subramanian, Karti Subramanian, Kristin Sullivan, Moshik Temkin, Dustin Tingley, Mike Toffel, Ian Tosh, Charlotte Tuminelli, Rodrigo Wagner, Jim Waldo, Steve Walt, Lee Warren, Rob Wilkinson, Julie Wilson, Carolyn Wood, Michael Woolcock, Josh Yardley, Andrés Zahler, David Zavaleta, and Pete Zimmerman.

At a more personal level, several people have influenced who I am as a human being, which inevitably has shaped who I am as a teacher: my wife Gaby, my two daughters Dani and Alex, my siblings Vanessa and Ari, and many other members of my family. My parents John and Licita deserve special mention as they embody many of the values that I aspire to have as a teacher, including empathy, generosity, kindness, and respect for every human being.

Finally, I would like to thank the many students I have had at the Kennedy School. Wherever you are in the world, thank you for the honor of having been in your path and for inspiring me day in and day out to bring the best in me to the noble profession of teaching.

About the Author

David Franklin is a British writer who is fascinated by how we learn about statistics. After taking Dan Levy's famous statistics class in the fall of 2018, he returned as part of the teaching team the following year and took notes on everything Dan did: those notes became *Invisible Learning*.

He has two master's degrees: one in mathematics from the University of Cambridge, the other in development economics from the Harvard Kennedy School. He also has eight years of experience in the private sector as a country risk manager.

In his spare time, David is an avid reader, radio cricket commentator, and long-suffering fan of Newcastle United. He lives in London with his partner Catri.

REFERENCES

[i] Clark, R v [2003] EWCA Crim 1020 (11 April 2003), paragraph 99

[ii] Eric Johnson and Daniel Goldstein, 2009. *Defaults and Donation Decisions.* Transplantation, Vol. 78, No. 12, pp. 1713-1716

[iii] Eric Hanushek, 2011. *The Economic Value of Higher Teacher Quality.* Economics of Education Review, Elsevier, vol.30(3), pp. 466-479

[iv] Ronald Heifetz, 1994. *Leadership Without Easy Answers.* Harvard University Press, pp. 8

[v] Charles Dickens, 1854. *Hard Times.* Bradbury & Evans, pp. 1

[vi] Daniel Willingham, 2009. *Why Don't Students Like School? A Cognitive Scientist Answers Questions About How the Mind Works and what it Means for the Classroom.* Jossey-Bass, pp. 19

[vii] George A. Miller, 1956. *The Magical Number Seven, Plus or Minus Two: Some Limits on our Capacity for Processing Information.* Psychological Review, Vol. 101, No. 2, pp. 343-352

[viii] Ronald Heifetz, 1994. *Leadership Without Easy Answers.* Harvard University Press, pp. 24

[ix] Ibid, pp. 49

[x] Ibid, pp. 38

[xi] Ibid, pp. 252

[xii] Ibid, pp. 238

[xiii] Louis Deslauriers et al, 2019. *Measuring actual learning versus feeling of learning in response to being actively engaged in the classroom.* PNAS 116(39) pp. 19251-19257.

[xiv] Ibid, pp. 19251

[xv] Ronald Heifetz, 1994. *Leadership Without Easy Answers.* Harvard University Press, pp. 104

[xvi] Louis Deslauriers et al, 2019. *Measuring actual learning versus feeling of learning in response to being actively engaged in the classroom.* PNAS 116(39) pp. 19256.

[xvii] Ronald Heifetz, 1994. *Leadership Without Easy Answers.* Harvard University Press, pp. 105-106

[xviii] James Lang, 2016. *Small Teaching: Everyday Lessons from the Science of Learning.* Jossey-Bass, pp. 28

[xix] Horace Mann, 1848. *Twelfth Annual Report to the Secretary of the Massachusetts Board of Education.*

[xx] Thomas Merton, 1979 (posthumous). *Love and Living.* Farrar, Straus Giroux. Edited by Naomi Burton Stone and Brother Patrick Hart.

[xxi] Mary Budd Rowe, 1972. *Wait-Time and Rewards as Instructional Variables: Their Influence on Language, Logic, and Fate Control.*

[xxii] Julia Rozovsky, 2015. *The five keys to a successful Google team.* Google re:Work online.

[xxiii] Amy Edmondson, 2019. *The fearless organization: Creating Psychological Safety in the Workplace for Learning, Innovation and Growth.* John Wiley & Sons, pp. 10

[xxiv] Ibid, pp. 14

[xxv] Ibid, pp. 35

[xxvi] Julia Rozovsky, 2015. *The five keys to a successful Google team.* Google re:Work online.

[xxvii] Ken Thomas, 2019. *Biden, Warren Gain in Latest Poll of Democratic Primary Voters.* Wall Street Journal online.

[xxviii] Aidan Colville et al, 2020. *Enforcing Payment for Water and Sanitation Services in Nairobi's Slums.* NBER Working Paper No. 27569.

[xxix] Elizabeth Word et al, 1991. *The State of Tennessee's Student/Teacher Achievement Ratio (STAR) Project.* Tennessee State Department of Education.

[xxx] Ronald Heifetz, 1994. *Leadership Without Easy Answers.* Harvard University Press, pp. 4

[xxxi] Ibid, pp. 76-88

[xxxii] Ibid, pp. 86

[xxxiii] Ibid, pp. 85

[xxxiv] Robert Fritz, 1991. *Creating.* Ballantine Books, pp. 41-64

[xxxv] Ronald Heifetz, 1994. *Leadership Without Easy Answers.* Harvard University Press, pp. 128

Printed in Great Britain
by Amazon

60502469R00194